To my Dear Friend

Glad to be part of the cause of Educating Patient's mind.

Best of luck.

FREEDOM FROM CPAP:
SLEEP APNEA HURTS,
the Cure Doesn't Have To

David Dillard and Mayoor Patel

ISBN: 978-1-4834-2383-8 (sc)
ISBN: 978-1-4834-2382-1 (e)

Library of Congress Control Number: 2014922724

Lulu Publishing Services rev. date: 1/15/2014

Contents

Acknowledgments

The tragedy of life is not that we die, but rather
what dies inside a man while he lives.

—Albert Schweitzer

As doctors, we are humbled and encouraged by the words of this great humanitarian, philosopher, and physician. Inspired by Dr. Schweitzer's assertion, our oath to "first do no harm" takes on additional significance when dealing with the diagnosis and treatment of sleep apnea and fundamentally has motivated the writing of this book. As we work to deliver individualized therapeutic solutions, our goal is to provide our patients the freedom and renewed zeal for life that overcoming sleep apnea both embodies and enables.

We would like to thank our families and the staff of both the Sleep and Sinus Centers of Georgia and the Craniofacial Pain & Dental Sleep Center of Georgia for their constant support as we toiled to develop the narrative; we are doctors, not writers. A special note of gratitude is long overdue for our wives, Teresa Dillard and Mitu Patel, who for years gracefully withstood both our training and our snoring! Last but not least, we would like to acknowledge the creative work of Sara Berg and Michael Hagan for their help with compiling, editing, designing, and producing this first collaborative effort.

David G. Dillard, MD
Mayoor Patel, DDS, MS

Introduction

Without sleep, we all become tall two-year-olds.

—Jo-Jo Jensen, *Dirt Farmer Wisdom*

An insight handed down by a Depression-era subsistence farmer to his granddaughter, who celebrates his wit and simple philosophy of life through her prose—the above nugget of wisdom speaks directly to the sleep issues that this book seeks to address. Tens of millions of Americans suffer directly or indirectly the ill effects of sleep apnea. They suffer unnecessarily. This book provides hope! The vast majority of those suffering could be helped through virtually pain-free intervention. The need is great. The awareness is lacking. Inadequate sleep takes its toll on the body in both apparent and more subtle, insidious, and life-threatening ways. This book will first outline the risks and then hammer home the reality that you simply do not have to live with sleep apnea. Simple, noninvasive, and nonsurgical solutions exist that can break the cycle of sleep disturbance and oxygen loss in even the most severe cases.

Our iconic subsistence farmer probably never saw a doctor in his life. Why then do we hold him up as an example of how to address sleep disorders? Indeed, the dirt farmer's plainspoken truth defines a treatment philosophy fundamental to our clinical approach: medicine isn't magic; it's logic.

We are David G. Dillard, MD, of Sleep and Sinus Centers of Georgia, and Mayoor Patel, DDS, MS, of Craniofacial Pain & Dental Sleep Center of Georgia. If you or a loved one suffers from sleep apnea, this book was written for you. Our objective is to provide information about new treatment options designed to ensure that patients experience the restorative sleep necessary for good health and vitality. A sleep specialist can assess your individual situation and provide resources that will empower you to better educate yourself concerning all the options available. Our team is dedicated to helping you achieve the result—the restorative sleep—that has been eluding you despite your current treatment regimen. We are excited about these new techniques. They are nothing short of revolutionary. Implementation has translated to better outcomes with fewer complications, less downtime, and demonstrably greater patient comfort. Over the past ten years, we have witnessed remarkable advances in the diagnosis and treatment of sleep apnea. Please, for your own health, and for the sake of those you love, talk to your doctor about these life-changing treatment options. Don't settle for the unfortunate alternative. Advocate for your health. Keep an open mind to advances in medicine. It's just common sense.

Common sense medicine: that's what this book is all about. There's a killer out there: obstructive sleep apnea. There is a pervasive ignorance of its impact on lifestyle—even on life itself. The disease is a factor in over 10,000 deaths annually. But don't despair. This book explores procedures and lifestyle guidelines developed to prevent, reduce, or even cure the disease that is slowly robbing you or a loved one of all the good things in life. Simply put, this book just might save your life. Common sense and our common concern for your health and safety urge you to keep reading.

Having picked up this book, chances are that you are in a long-term relationship and snoring has driven a wedge between you and the person with whom you share (or no longer share!) a bed. That wedge has an undeniable emotional cost. Physically, you may appear perfectly fine, or you may be constantly tired and feel ill. Regardless of how you look or feel, someone has seen you snoring or gasping for breath at night, or you have some medical condition associated with sleep apnea that caused your doctor to order a test. This is a call to action. The problem will most certainly not go away on its own. Delaying treatment may place further strain on your relationship and will ultimately damage your health.

Perhaps you are reading this book because you have already sought treatment for snoring or obstructive sleep apnea (OSA) but are discouraged and frustrated because the "solution" is either not working or is too cumbersome and not a good fit for you lifestyle.

Whatever the case, we are happy that you have taken the first step. By arming yourself with information, you will discover that there really is hope. We started this book because we want to get across one message: don't give up! The risk—no, the *danger*—of inaction when symptoms of sleep apnea are present is too great to take a wait-and-see approach. Once sleep apnea has been diagnosed and a treatment plan is put into action, failure is no longer an option. *Sleep apnea will not improve without treatment.* In light of recent advances in both surgical and nonsurgical techniques for the treatment of sleep apnea, giving up on one and not seeking a better alternative is nothing short of foolish and might ultimately prove deadly.

We practice mainly in the field of sleep medicine, and most of our patients arrive in our office desperate for help. We are their last line of defense. One of the most common responses we hear from patients is that they feel defeated. They cannot tolerate the CPAP mask and have been unable to lose weight. One of the most gratifying things we consistently see is that our clinics restores patient optimism. Indeed, rest assured! An effective, manageable treatment exists that will free you from sleep apnea's "obstruction." Please, don't give up! The right treatment *will* restore your physical and emotional well-being and in all likelihood extend your life.

We've all heard the saying "The treatment is worse than the cure." When it comes to sleep apnea, nothing could be farther from the truth. So let's table that well-worn phrase. Consider instead "Life is good." Given that restorative sleep with steady oxygen delivery to the brain is an intrinsic good, the message resonates. So seek treatment for OSA, and live *well*.

Left unchecked, sleep apnea will slowly rob you of your health and mental acuity. Ironically, masked by the often high-decibel snoring that can accompany it, sleep apnea is indeed a silent killer. As we will outline in later chapters, health risks include high blood pressure, heart arrhythmia, metabolic problems, diabetes, cognitive loss, memory issues, obesity, erectile dysfunction, stroke, heart attack, and a host of other serious ailments.

Here's the good news: intensified research conducted in recent years targeting the virtual epidemic of sleep apnea has paid off. Yesterday's overtly intrusive, painful, and regrettably ineffective surgical solutions have been replaced with highly effective options that mediate the symptoms of sleep apnea. These new technologies afford patients treatment options that won't interfere with their busy schedules while improving—not compromising—quality of life.

Sleep apnea hurts. It hurts productivity, relationships, health, and well-being. Freedom from CPAP means that the cure doesn't have to. There really is no reason to avoid or even delay treatment. As our friend behind the plow might say, "Stop cussing the mule, and load the wagon."

PART I

Diagnosis

Chapter 1

Why Do We Need Sleep?

A good laugh and a long sleep are the
best cures in the doctor's book.

—Irish proverb

Deadlines, schedule conflicts, childcare, paying bills—life can be challenging, even on a good day! If you are like most everyday heroes scrambling to meet the merciless demands of an impossible schedule, you might contemplate skipping sleep. It won't do any harm, right? Wrong. Losing even a minimal amount of sleep compromises mood, energy, and the ability to handle stress. Understanding the importance of sleep and why you need proper rest to maintain your daily schedule is the first step in adopting, maintaining, and prioritizing the sleep schedule necessary to protect your overall health. In fact, *you can't afford to lose sleep.*

Sleep is as critical to life as food and water. Adequate sleep every night enhances alertness, energy, happiness, and both cognitive and motor skills. It helps us to feel our very best. Conversely, going without sufficient sleep causes drowsiness, lethargy, depression, diminished mental acuity, and well, clumsiness. Such observable, and therefore measurable, negative outcomes, perpetuated by habitual sleep deprivation, are just the tip of the iceberg.

To help you better understand the importance of sleep, let's look at what happens when you experience a sleep deficit. If you have ever pulled an all-nighter, you probably understand just how challenging—even impossible—it was to function at your best the next day. While missing one night of sleep is not fatal, it can be detrimental to your health.

Losing a night of sleep can cause you to feel irritable throughout the next day. It can also cause you to slow down and become tired easily. However, for some individuals, it can also cause you to become wired by adrenaline. All-nighters produce a drowsy, punchy, or simply overly caffeinated version of the real you. Only self-deception might convince you that you are at the top of your game when you go without sleep!

If you lose two nights of sleep, your performance declines more sharply. Concentration becomes difficult—even sporadic—and your attention span becomes squirrel-like. (No offense to the furry critters, but they do distract easily.) Mistakes increase, requiring more time to complete daily tasks. Punchiness gives way to edginess and irritability, and ultimately emotional fragility or breakdown. The more sleep you lose, the more your mind and body begin to shut down. After a third day of no sleep, you might begin to hallucinate, which can cause clear thinking to diminish entirely. While these examples might appear extreme to some of us, you can still experience many of the same problems over time if you sleep only a few hours per night. As the sleep deprivation increases, the more your symptoms will worsen. In time, a lack of sleep can actually be fatal.

Just as exercise and nutrition are essential for optimal health and happiness, so too is sleep. Quality, restorative sleep every night directly affects your life in number of critical areas, including

- mental sharpness
- productivity

- emotional balance
- creativity
- physical vitality
- weight

Sleep is actually quite analogous to eating. When you are hungry, you eat. When tired, you rest. (At least you should.) By eating, you are relieving hunger and ensuring that your body obtains the nutrients necessary to sustain good health. When you sleep, you are relieving sleepiness and ensuring that you obtain the sleep you need to function throughout the day. Still, many people seem to question the need for sleep and try to minimize sleep with stimulants, supplements, and other inadequate alternatives to the real thing!

What then makes restorative sleep unique? Let's call it *the power of sleep*. While many of us try to sleep as little as possible, it is important to pay attention to why just a few more hours of sleep can have a tremendous impact on our health and well-being.

Contrary to popular assumptions, sleep is not a time for your brain and body to shut down. Rather, while you rest, the brain continues to work, overseeing an array of vital biological maintenance processes focused on keeping the body running and in preparation for the next day. However, as mentioned before, restorative sleep cannot occur on a truncated or interrupted schedule. The brain is unable to complete its maintenance and regulation of the body's complex chemistry. As a result, work, creativity, learning, and communication suffer. Inadequate sleep prevents us from performing to our full potential. To help you get a better night's sleep, the National Sleep Foundation suggests that you practice the following:

- Go to sleep and wake up at the same time every day.
- Avoid spending more time in bed than needed. (Reserve your bed for sleep.)
- Use bright light to manage your body clock.

- Select a relaxing bedtime ritual.
- Create an environment that will help you sleep—typically quiet, dark, and cool.
- Reduce your intake of caffeine, or eliminate it completely.
- Keep bedtime a worry-free time.
- Exercise regularly, but avoid vigorous workouts before bedtime.

Chapter 2

Snoring

People who snore always fall asleep first.

—Unknown

This humorous saying provides a fitting entrée to our discussion of snoring—unless of course you are the one subsequently unable to fall asleep. Ironically, there may be more truth to the saying than the obvious conclusion: one snores only when one can be heard by another (restless) soul! For instance, it is likely that the snorer suffers from undiagnosed sleep apnea and is thus deprived of restorative sleep on a regular basis. As a result, he is more tired and therefore falls asleep faster than his audibly frustrated bed partner.

There is indeed a link between obstructive sleep apnea and snoring. However, both the disease and the phenomena can occur independently. That is to say all who snore do not have sleep apnea. It is important to also note that that the absence of snoring does not indicate freedom from the disorder.

While snoring may seem like a relatively simple matter, with ongoing progression, it can have unfavorable effects on a person's health and relationships. A narrowing of your airway causes snoring. When there is a partial obstruction to the free flow of air through

7

your mouth and nose, this causes a decrease in airflow to the lungs and a lack of oxygen to the brain. A narrow airway hinders smooth breathing, which in turn creates the sound of snoring and can be disruptive to you and those nearby.

If you snore, you are not alone. In fact, almost half of the population is affected by snoring to some degree. When it becomes a nightly occurrence, however, it can have detrimental effects on your health and relationships. Fortunately, treatments are available that can help open up your airway for unobstructed breathing, allowing you and those around you to achieve a better night's rest. But we will cover that in later chapters.

When we think of snoring, we immediately think of the annoyance it causes for our sleeping partners. Snoring can have many negative effects on your life and overall well-being. Unfortunately, snoring and sleep go hand in hand, which can have a negative effect on your health, your personal relationships, and your overall quality of life. If you snore, you may wake frequently from the snoring or the constant tossing and turning. This can make sleeping difficult for your spouse or partner as well. The ongoing sleep disruption and excessive tiredness resulting from snoring can strain even the most dedicated relationships, leading to resentment, sleeping in separate rooms, and in some cases, divorce.

Severe cases of snoring can cause serious long-term health problems as well, including potentially life-threatening obstructive sleep apnea, which has been linked to heart failure, high blood pressure, and stroke. A person with sleep apnea will wake up several times throughout a night in order to regain their breathing that is continuously being disrupted. Additionally, irritability, sleepiness, and lack of productivity during the day are other serious consequences of excessive sleep deprivation.

Snoring management may include self-help remedies, including

- sleeping on your side
- losing weight
- limiting alcohol and medications before bed
- elevating your head during sleep

Eliminating your snoring can result in a better night's sleep for both you and those around you, improvement of your health, and ultimately a boost to your overall quality of life. If you are tired of losing sleep and want to get the rest you need to complete your daily tasks, educate yourself.

The National Sleep Foundation's 2002 Sleep in America poll revealed that 37 percent of adults report they had snored at least a few nights a week during the previous year. In fact, 27 percent said that they snore every night or almost every night. Males were more likely than females to report snoring at least a few nights a week (42 percent vs. 31 percent).

To the extent that snoring is the number-one risk factor for obstructive sleep apnea, please consider taking the quiz developed to help quantify your sleep deprivation, regardless of cause, and qualify the need to seek treatment. Even if you don't snore (or you might just be the second one to fall asleep!), we urge you to take the quiz located on the following page.

The Epworth Sleepiness Scale

The Epworth Sleepiness Scale contains a list of eight situations in which you will rate your tendency to become sleepy. The test is on a scale of 0 (no chance of dozing) to 3 (a high chance of dozing). Your total score is then added up and evaluated on a scale of 0 to 24 to determine your level of sleepiness.

Situation	Chance of Dozing
Sitting and reading	
Watching Television	
Sitting inactive in a public place (e.g. a theater or a meeting)	
As a passenger in a car for an hour without a break	
Lying down to rest in the afternoon when circumstances permit	
Sitting and talking to someone	
Sitting quietly after a lunch without alcohol	
In a car, while stopped for a few minutes in traffic	
Total score (add the scores up) to determine your Epworth score.	

A score greater than 9 indicates a possibly acute condition that should be looked at by an MD specializing in sleep medicine. It is important to note that individuals who score low on the Epworth Sleepiness Scale might still have a sleep disorder that should be evaluated by a doctor.

Chapter 3

The Silent Killer

The killer awoke before dawn ...

—Jim Morrison

One of the most common and vivid nightmares a person can have is of being smothered in his or her sleep. Being strangled or smothered is such a primal fear that mystery and horror films employ it routinely. Drowning is a similar scenario. There is a common thread. It is a loss of control of the most basic of functions: breathing.

If there was a killer on the streets sneaking into people's homes and smothering them at night, it would make national headlines and there would be a huge manhunt. If a tobacco company was selling a product that affected roughly half the population at age fifty and caused as many heart, lung, and stroke problems as does sleep apnea, public charities would be formed that would dwarf the combined influence of both the American Lung Association and the American

Heart Association. The fact that sleep apnea alone causes a sevenfold risk of motor vehicle accidents should have every Oprah Winfrey protégé campaigning to find the cure. *Truly, sleep apnea is an enormous problem that can no longer be ignored.*

Why is there not a larger outcry? One factor seems to involve a misunderstanding, even among physicians, regarding the root cause of sleep apnea. Sleep apnea may have a new name, but sleep problems have described for centuries. Charles Dickens' *Pickwick Papers* describe a red-faced, obese boy who six decades later would inspire the great physician Sir William Osler (who practiced at the turn of the last century) to name a disease Pickwickian syndrome. The general consensus since then has been that the patient's snoring, and thus also sleep apnea, is due to obesity and gluttony.

The following narrative explores Dr. Dillard's personal journey with sleep apnea and demonstrates how his perception of the disease was influenced in a direct way by his own diagnosis:

> The fallacy of the common misconceptions concerning sleep apnea didn't dawn on me until I came to realize and accept that I myself was suffering from the disease. At the time, I was in the best shape of my life. I was running marathons—twenty-six miles' worth of running—and thought, *You must be kidding!* Then, as I started looking into it, I noticed other people in my practice with sleep apnea that didn't fit the stereotype. The buff twenty-year-old Marine with normal tonsils and nasal obstruction was a paradigm shift for me. This young man was about to be removed from active duty, despite an obviously promising career in the marines. He was not allowed to remain in his position with a CPAP machine. (He couldn't very well take a CPAP into a foxhole.) That is when I really started to question what sleep apnea really was all about.

Clearly it is not just obesity, as our buff marine with the six-pack abs and 8 percent body fat falls far outside that category. There must be something else. So let's take a closer look at what sleep apnea is. Let's start with the two main types of sleep apnea: central and obstructive. The third type is a combination of the first two types and is called mixed, or sometimes *complex*, sleep apnea. (Really, as it is both uncommon and will not affect your basic understanding of the disorder, we will not discuss mixed sleep apnea.)

Central sleep apnea is lack of breathing that comes from a loss of drive from the brain—part of the central nervous system—to breathe. Your brain kind of forgets to move your diaphragm. Now this form of apnea (*a* means "no" and *pnea* means "breathe") is much less common than the obstructive variety. It accounts for less than 5 percent and is associated with various forms of neurologic disorders like stroke and developmental delay in children, or serious medical conditions that affect oxygen and carbon dioxide exchange. Examples of this kind of problem would be someone with serious heart failure or lung disease that doesn't get the usual amount of oxygen to the brain and does not get rid of the waste product of burning food (carbon dioxide). When these two problems happen to the fuel supply in the brain, there is a decreased drive to breathe on the usual cues, causing an individual to hold his or her breath. Central sleep apnea needs an artificial drive to fix this problem, a sort of artificial ventilator, if you will.

Because of the often close association between neurological disorders and pulmonary problems, two of the pioneering specialties managing central sleep apnea are lung doctors and neurologists. These specialists are most frequently practicing both pulmonary or neurology and sleep medicine at the same time. Fortunately, central sleep apnea diagnoses usually makes up less than 10 percent of any given sample of sleep apnea cases.

The second type of sleep apnea is much more prevalent: *obstructive sleep apnea,* commonly referred to by its acronym OSA. The name is fairly descriptive and accurate: there is an obstruction of the breathing at some point. Instruction in the anatomy of the tongue and throat would aid in developing an understanding of what is going on. But first let's introduce some common sense. I was going to say physics but figured your eyes would glaze over as you might imagine sleeping in the back of some high school or college classroom lecture. So just relax, put the calculator away, and consider a garden hose. Everyone has had to turn on a garden hose. You have a length of rigid or semi-rigid pipe comprising the plumbing fixture with a valve handle attached to it. To get more water, you open the valve to release pressure from the rigid pipe which then flows into the hose line.

Everyone I know has used a garden hose. The physics involved in the process of obstructive sleep is just about as simple. Let's break down the physics. Because the continuous pressure supplying water to the fixture is greater than the environment outside the fixture, opening its valve (lefty loosey!) will increase water pressure in the attached hose, causing more water to flow into or through the hose whether it is kinked or not. Water flows downhill (meaning it follows the path of least resistance) and comes out when you open valve, tap, spigot, faucet, or whatever you like to call the thing.

Now breathing is a lot like a hose in a many ways. The diaphragm is a muscle that creates a pressure wave of air. It expands the volume of the lungs and creates a pressure difference between the air in the outside world and the lungs. So I guess it is kind of like a vacuum cleaner as well in some respects, but I think you get the picture. When you breathe in, the outside air is under more pressure than the air in your lungs and it flows into the lungs. When you breathe out, the pressure is increased in the chest. In the obstructive type of sleep apnea, the airway is narrowed, at least temporarily, and it kinks or pinches the flow. The same is true when you kink a hose: you can actually stop the flow, at least temporarily.

Clearly, in the case of the breathing example, kinking the airway has more dire consequences than kinking the hose to the kids' blow-up pool, but the physics are similar. What is actually happening is that you are narrowing the hose by kinking it. I know this is intended to be a nonscientific illustration, but let's insert some mathematics now for a minute—but only because it's important! When you double the size of a pipe carrying water, air, or any other fluid, the amount that goes through is greater. That's obvious. What is not obvious is that the flow doesn't just double. For any given pressure, if a pipe's radius (half the diameter, in case you've forgotten) goes from 1 to 2, that doubling in size increases flow 8x (2x2x2). Tripling the radius from 1 to 3 increased the flow by a factor of 27 (3x3x3). That is an exponential increase in flow. Put another way, the flow is increased to the third power or exponent of the radius. (On second thought, maybe the numerical expression of the factors is easier to comprehend than the third exponent of the radius ...)

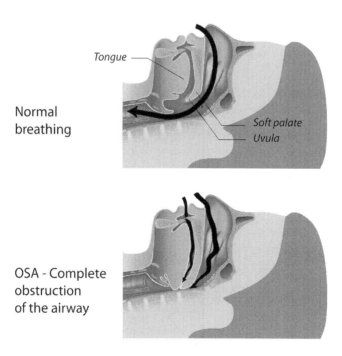

Tongue

Normal breathing

Soft palate

Uvula

OSA - Complete obstruction of the airway

Wider Is Way Better

Now that everyone's eyes have glazed over, let's make it really simple. When it comes to the airway and oxygen delivery, wider is way better. This concept is important only because small changes in the diameter of the throat and nose can have a big impact on airflow, *a very big impact*. We will come back to this concept again later. The obvious solution in the garden hose example is to unroll and unkink the hose. Easy enough. That said, it might seem easy enough in principle for the throat, but the complexity of the throat and nose make the analogy more challenging to address from a treatment standpoint.

"Unrolling the hose" once only meant surgery. In the past, it meant major surgery. Unfortunately, highly invasive efforts to "open the hose" permanently compromised some other important things that the throat and nose do a very good job of (swallowing, for example). For now it's enough to get the idea that small changes in the size of the pipe are *very* important.

Now there are some instances in which surgical procedures can improve the airway with few or no negative side effects. Tonsils, for example, are generally structures that can be removed with little side effect other than pain—except in very rare instances. We will get back to this and other surgeries later as well.

Now looking for a way to help these folks who stop breathing at night, someone early on in sleep medicine got the bright idea that, if you turn up the pressure enough, it will expand the throat and hold it open to breathe, and air *will* flow. Thus, the advent of the CPAP (continuous positive airway pressure) machine.

The following illustration is not exactly what happens with CPAP, but if you have ever seen a kink in a hose unwind with increased pressure, then you understand the basics of CPAP. CPAP is actually more akin to one of those collapsible hoses (like a fire hoses or, as an extreme example, one of those kink-free hoses from late-night infomercials). It takes a certain amount of pressure just to expand the hose and keep it open. Once the hose expands, the water flows freely.

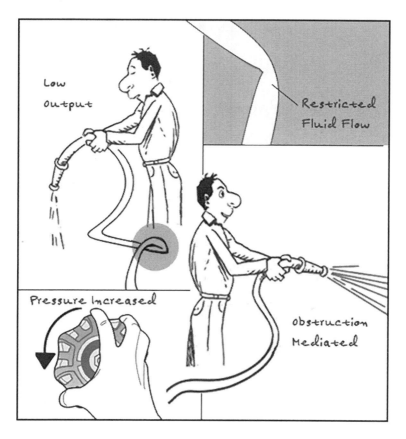

The Face Is Not a Faucet

The gist of the problem is that it has been very difficult in the past for some patients to get air into the nose while using CPAP. Like a hose that is not screwed on tightly or lacks a rubber washer, the CPAP, for a lot of people, will leak.

Water and air have some similarities. They are both fluids. They fill any container you put them in, and they flow. Water is just a lot denser. Also, the face does not generally come with a threaded attachment or quick-release coupling for the insertion of pressure hoses. In short, the face is not a faucet (or a spigot, to use the vernacular). To keep the pressure going in, the attachment to the face uses a pressure mask to drive the air into the throat via the nasal passages. However, the only way yet to put more pressure into the airway is to put more pressure on the face, lest the air pressure takes the path of least resistance and leaks out at the seal.

If the pressure of the mask on the face is not greater than the pressure in the CPAP, the CPAP will indeed leak. This increased pressure alone bothers most people, particularly when addressing more severe cases of apnea. As therapeutically necessary pressures rise, the irritation on the face increases. This is particularly true over the bridge of the nose. I have seen pressure sores on the nose caused by the CPAP. Higher pressure also tends to dry out the nose, which tends to cause nosebleeds.

Aside from not doing its job to pressurize the airway, air leakage itself can pose a problem. The leakage bothers people's eyes, and the drying effect can potentially cause or exacerbate serious problems with the eyes. More often, however, the noise from the leak bothers the patient or their partner. Again, more importantly, even though advanced CPAP machines can now compensate for the leak, the residual pressure will fundamentally not do an optimal job of pushing the throat open and enhancing airflow into the lungs in the presence of a leak.

The act of smothering (referring again to the metaphor of the silent killer) cuts off the supply of oxygen, causing death by asphyxiation. That is ultimately what is happening with the patient who either doesn't use the CPAP properly or who cannot tolerate it. Fortunately, through the recent integration of internal computers, the masks have gotten better and the machines smarter. The newer machines can detect when there is a leak and when you are breathing. They can detect when more pressure is required to open your airway and then adjust the pressure accordingly, and automatically. The newer machine can tell your doctor how often, and even how effectively, you are using the machine. These machines have better humidifiers and are quieter. The masks are better at distributing the pressure evenly, which results in improved patient comfort. So CPAP is much better than it used to be and can be of immense benefit.

Here is the key difference to consider: once a procedure is performed, you can't take off the surgery. Conversely, you can take off the mask. That simple truth is a huge part of our philosophy. We always try the least invasive option first. You have time because sleep apnea is a disease that progresses slowly. It is better to first try everything that you can take off or take out or take back before performing even a minimally invasive surgical technique. Moreover, the data than can be collected while "under the mask" provides critical information to physicians concerning the specific nature of your sleep apnea and is invaluable in guiding decisions regarding subsequent procedures that may be deemed necessary based on a quantitative analysis of the severity of your symptoms.

The pressure and claustrophobia are the number main reasons people cannot handle the CPAP mask. While it is reasonable to assume that you picked up this book because you want to get relief from the mask, it is important to keep things in perspective. Let's be realistic. The risks of modern sleep apnea surgery or dental devices (discussed later in the book) are minimal but present. You can take off the CPAP mask and you can take out the dental device. The CPAP has been proven effective for some people, and it's worth a try. Also, your insurance carrier or third party payer is usually going to require you to try the CPAP first in any case. Therefore, let's talk about some things that can help you manage your CPAP better.

First, it's important that your sleep technician or registered respiratory therapist allows you to choose from the several mask styles currently offered by the handful of CPAP manufacturers who produce and distribute the machines. In this case, variety is not only the spice of life but also the broker of life-giving oxygen to abate the threat of an intimidating array of heart, lung, and brain problems that stem from untreated sleep apnea.

One of the common reasons why masks fail is because they ride up on the eyes. Chinstraps can be helpful, although proper fitting with by a trained professional is advised. Oftentimes, there is a tendency for the mouth to open up at night, allowing air to escape. This tendency can cause problems that, again, a properly fitted chinstrap can help to mediate. With so much variation in human facial contour, it is wise to try as many options as possible before selecting the mask that affords you the most comfort. The masks all differ in fit and feel. You owe it to yourself to try them all.

Some patients find that the mask falls off at night because they change position. There are specially made pillows to help with this. These "CPAP pillows" or "contour CPAP pillows" are sold both online and at a growing number or department stores, bedding retailers, medical supply centers, and even Bed Bath & Beyond!

There are often times that the mask collapses the side walls of the nose (especially in patients with nasal valve collapse) and nasal dilation devices such as breathe rite strips can help. With several distinct products on the market designed to enhance airflow through the nasal passages, this partial list will help you find and try the product that best addresses your particular nasal issue:

- Eden 3000 "Silent Sleep—Snore Stopper"
- Flents™ Breathe Well Nasal Dilator
- Snorepin®
- Sleep Right Nasal Breathe Aids
- RespiFacile®

A third option is to order nasal pillows, which is actually a CPAP attachment that goes in your nose. This can help with nasal valve collapse. While many options are available, retail availability is primarily limited to medical supply centers or online. Here are a few common styles and brands to consider in your search:

- Swift™ FX Nasal Pillow
- AirFit™ P10 for Her Nasal Pillow CPAP Mask
- Opus 360
- OptiLife Nasal Pillow

Beyond pressure sensitivity and structural limitations that contribute to leaking around the mask, some people who use CPAP are claustrophobic and feel like they are choking. Should you find this to be the case, try using the mask in conjunction with a sleep aid or prescription for medications to reduce anxiety (anxiolytic). Common sleep aids include over-the-counter preparations like Tylenol PM or Advil PM. Prescription sleep aids include drugs like Ambien or Lunesta. All sleep aids should be taken with caution. Patients with sleep apnea that is untreated need to be particularly cautious. Make sure that your doctor or provider is aware that you have sleep apnea if you ask for these medications. Medications prescribed to reduce anxiety include Valium and Xanax. These drugs can be helpful in training the patient to tolerate and eventually make peace with the CPAP mask. It is important to note that narcotic pain medication to get to sleep is not a good idea because people tend to develop a tolerance to the medications. I am generally talking about opioid medications like codeine, hydrocodone, oxycodone, or morphine-like medications. People require more and more medicine to get to sleep, and the medicine can actually stop you from breathing. That is a real problem if you already have sleep apnea, or suffer from heart or lung problems. All of these medications need to be used under close medical supervision.

Sometimes, starting out with the mask on for a long time before you go to bed helps with the claustrophobia. Sometimes, sitting more upright in bed also helps. There is a study that states that Afrin spray used at a very small dose only once a day does not develop tolerance and can help keep the nose open. Such a therapeutic use (again, in very small dose) differs from the "Afrin test" that is promoted to help diagnose chronic sinusitis. Again, check with your doctor before starting any course of medication.

A small amount of skin emollient, cream, or even petroleum jelly on the skin can also help stop the leaking and skin irritation. This can be most helpful over the bridge of the nose, around the tip of the nose, and around the mouth.

If all else fails, there are also CPAP clinics dedicated to helping patients adapt to the device, resulting in markedly increased success rates. These visits are often covered by insurance providers, which recognize how damaging sleep apnea is to your heart and want you to use the machine.

Long-term CPAP use can cause sinus problems and nose bleeding. While a major cause of CPAP failure, sinus problems persist undetected in 30 percent of patients who engage our practice because they feel that they can no longer tolerate the mask. The sinus problems are diagnosed via a CAT scan of the sinuses, which is nearly 100 percent accurate in identifying sinus problems, even if the patient is asymptomatic.

Now it may surprise you that sinus problems can cause no symptoms. Most of us have had the experience of having a sinus infection associated with a cold. We think of it as a pressure around the eyes and discolored drainage. We also think of a sinus infection as causing problems breathing through the nose. Some people call this congestion, but we all know that we don't breathe well with a cold, which is a type of viral sinus infection.

But if the cold lasted all your life, how would you know what normal breathing was like? You would have no frame of reference.

We kind of forget physical feelings. One of the things I find really interesting is that people with severe nasal problems get inflammatory growths called polyps in the nose. Now most of us have heard about polyps in other areas of the body that are precancerous growths. These are not. They develop from chronic swelling in the nose and the sinuses. They can literally close the nose off so that these people have no option but to breathe through the mouth.

What is even more amazing is that patients sometimes don't even recognize that they have a sinus problem. I will look into their nose and see large, inflamed polyps and ask if they have problems breathing. Sometimes, the answer is, amazingly, no. Their CAT scans shows massive amounts of polyps, but sometimes their main complaint is that they can't smell.

Polyps develop gradually, until one day you notice their negative impact on your ability to breathe through the nose. It is like a wrinkle that you notice in the mirror and say, "When did that show up?" Sometimes, patients are so used to their breathing being bad that they have lost all frame of reference.

Here's another illustration: As folks age and don't sleep well, they will often just gradually get used to it. Feeling bad, in many cases a result of inadequate sleep, becomes the new "normal." Like the sinus patient in the preceding example, they lose all frame of reference and are unable to determine that their sleep problem even exists. It does not have to be that way. In such cases, sleep apnea is a great deceiver. It slowly takes away alertness and vitality and fools sufferers into the false belief that their fatigue and lethargy merely signal the loss of their last vestiges of youth.

There Is Hope

All this talk of polyps, poor sleep, and resignation to aging is included here to underscore our main point. *You do not have to live with the symptoms that are tearing you down.* The first step is to be evaluated and diagnosed by a sleep specialist, then take action to get your life back. Not only will the quality of your like improve in ways beyond your conditioned frame of reference, but you will also extend, if not save, your life.

Press On: Hope in Action

There is a motto used on navy ships that sums up the encouragement that we hope to provide through this book: "Press on." Despite all odds and all adversity, keep trying. The implication is that you will prevail. Oftentimes in life, the willingness to press on is ultimately what enables success. Tap into that spirit! It is important to keep trying to improve your sleep apnea. That's pressing on. That's hope in action. It is not, however, necessary to always go in the same direction. There is no single best way to treat the disease. Each individual must be a partner with his or her physician and must weigh the risk/reward of the various treatment options available. Like the fluid air itself that we breathe, together we will find the path of least resistance: the path right for you.

Too often when people are first told that they have sleep apnea, physicians—well-meaning physicians—will tell them to try harder. Try harder to lose weight. Try harder to keep the CPAP mask on. There is saying that doing something over and over again is the definition of insanity. Although you absolutely must give the CPAP a good solid try, this book is about finding another path if the mask won't work for you. If you come up against a wall and have to get past it, you have several choices. You can go over it, under it around it, or through it. Going through it may be necessary, but usually we try and take the easiest way. If you cannot get through using the CPAP, let's change course and consider some alternatives. There is a saying, "that doing the same thing over and over again and expecting different results is the definition of insanity."

Chapter 4

Environmental and Genetic Factors

The length of this document defends it well
against the risk of its being read.

—Winston Churchill

Sleep apnea is commonly associated with obesity and male gender. However, in reality, it affects a broad range of the population. A significant risk factor is habitual snoring; a recent study indicated that one in three men and nearly one in five women who habitually snore also suffer from some degree of obstructive sleep apnea. In addition to these statistics, we have found that ethnicity can also play a large role in the development of the disease.

Based on a National Sleep Foundation survey,[1] the findings suggest that African Americans face a higher risk for sleep apnea than any other ethnic group in the United States. Other groups that are at an increased risk for sleep apnea include Pacific Islanders and Mexicans. Data on Hispanics were limited in the study, but sleep apnea was nonetheless proven to be more common among this ethnicity when compared to its incidence among Caucasians. This also may be due to the differences in obesity between each population—a common factor leading to sleep apnea among all ethnicities. Similar to Hispanics, information about Asian Americans is also limited, but it is suggested that prevalence rates among these populations may be similar to Caucasians. While sleep apnea is not solely based on ethnicity, it is proven that it does play a key role.

When looking at risk factors for sleep disordered breathing, African American children were more likely than children of other races to develop obstructive sleep apnea. An increased risk of sleep apnea among African Americans is also independent of obesity or respiratory conditions as risk factors. A survey of symptoms among Caribbean-born black men and women found high rates of snoring (45 percent), excessive daytime sleepiness (33 percent), and difficulty maintaining sleep (34 percent).

Smoking and Sleep Apnea

Sleep apnea can affect anyone at any time, which is why it is important to be tested for sleep disorders even when you only display minor symptoms, such as snoring or perpetual daytime sleepiness. Again, anyone can suffer from sleep apnea, and that risk increases for individuals who smoke.

Smokers are three times more likely to suffer from obstructive sleep apnea than nonsmokers. Smoking is known to increase inflammation and fluid retention in the upper airway, further aggravating sleep apnea symptoms. In the end, both smoking and sleep apnea are deadly conditions that can severely shorten your life span. When combined, smoking and sleep apnea further increase the risk of cardiovascular and respiratory health problems, placing additional emphasis on the need to quit smoking while also seeking care for sleep apnea.

Protect yourself and kick the habit. Quit smoking now. Sleep apnea treatment is often a multistep process, which is especially true for smokers. Cigarette smoking increases inflammation in the upper airway and can aggravate symptoms, such as snoring and pauses in breathing (apnea). By addressing your smoking habit, you can improve your treatment options. While quitting smoking can't guarantee that your sleep apnea will disappear, it can ensure that the treatment option chosen will be much more effective.

Sleep Apnea in Pregnancy and Menopause

After passing menopause, if you find that you are nodding off more during the day, or during this book, it may be more than just maturity catching up with you. If your spouse or bed partner has reported that you snore or even stop breathing while you sleep, it could be sleep apnea. Women who are past menopause should be particularly aware of the symptoms of sleep apnea, as their risk of developing the disease increases. Moreover, when sleep apnea symptoms occur for these women, it poses a more severe risk than with younger women.

Menopausal women are one of the most at-risk groups of women for developing sleep apnea. However, many women who are going through menopause write off these symptoms as normal changes going on in their bodies. A woman might think her symptoms are normal and that once she gets through menopause the symptoms will get better. Her doctor might even agree, albeit this is clearly not the case. While not all symptoms are sleep apnea related, it is important to understand what to watch out for.

A woman might have sleep apnea if she feels run down, tired, and fatigued during the day. If she has a few moments to herself, she might feel like dozing off even though she had a good night's sleep. Other women might also notice that they wake themselves up feeling as if they are gasping for air and have no idea why they were suddenly jolted awake. Regardless, if you are going through menopause, do not ignore your symptoms—they are serious and should be evaluated by a sleep specialist. Seeking proper diagnosis and treatment is vital in maintaining your health during and after menopause.

During pregnancy, women who are otherwise quite healthy often develop certain complications. Undiagnosed sleep apnea may in fact be responsible for, and certainly contributes to, these problems. Pregnant women who suffer from untreated sleep apnea increase their risk of high blood pressure fourfold. Untreated sleep apnea doubles the risk of gestational diabetes. While it is commonly recognized that sleep apnea affects less than 1 percent of women of childbearing age, it is much more prevalent in those who are pregnant or have already gone through menopause.

Sleep apnea causes your airway to narrow and breathing to stop momentarily—sometimes up to hundreds of times a night. This lack of oxygen puts your body into a "fight or flight" mode, which in turn pumps out hormones like adrenaline and cortisol. When this occurs, it can send your blood pressure soaring. With sleep apnea, it also causes your body to produce more glucose so that there is plenty of energy to respond to this "threat," but over time, this can lead to diabetes.

By seeking treatment for sleep apnea, women may be able to prevent or manage complications during pregnancy. While this sounds scary, it is important to take heart because pregnancy-related sleep apnea typically resolves or improves after the baby is born. However, it is vital for your health and your babies that you receive proper treatment for your sleep-apnea symptoms immediately.

Genetics

There are many factors that can put someone at risk for sleep apnea. There is a strong possibility that these factors include genetics. If you have a serious condition like type-2 diabetes, high blood pressure, heart disease, or an endocrine or metabolic disorder, your chances of suffering from sleep apnea increase significantly. Smoking and alcohol use can also increase your risk of sleep apnea. A family history of sleep apnea should be discussed with your doctor. If you are experiencing loud snoring, lack of energy, daytime sleepiness, morning headaches, or depression, chances are you are suffering from sleep apnea.

Genetically inherited physical traits like your face and skull shape, characteristics of your upper airway muscles, and body fat content and distribution could all contribute to whether or not you are more prone to suffer from sleep apnea. Your height to weight ratio is often linked to sleep apnea, as obesity is a condition that is commonly associated with this condition. Again, it is important to remember that people who are fit can also suffer from sleep apnea.

According to the National Center on Sleep Disorders Research, if you suffer from sleep apnea, you may pass it on to your offspring, as it is not uncommon for members of the same family to suffer from the same form of sleep apnea. While this phenomenon has only begun to be researched, we can clearly see a link between genetics and sleep apnea.

Patients who suffer from obesity are more likely to suffer from sleep apnea. Since obesity is often caused by environmental factors, it stands to reason that members of the same family are more likely to be obese and then suffer from sleep apnea. Obesity affects the function of upper airway muscles, ventilator control, and conditions of sleep, which is why it is possible that obesity predisposes a person to sleep apnea.

Having a family member with sleep apnea can significantly increase your chance of developing or having this disorder and underscores the importance of understanding and being aware of the symptoms of sleep apnea. For individuals with a family history of sleep apnea, we strongly suggest taking a sleep quiz to determine your risk factors. In case you missed it, there is a sleep quiz located at end of chapter 2.

Down Syndrome and Pierre Robin Syndrome

Pierre Robin syndrome is a congenital condition of facial abnormalities in humans. It is a chain of certain developmental malformations, one entailing the next in a sequence. The three main features of this condition are cleft palate, micrognathia (small jaw), and glossoptosis (downward position of the tongue). Down syndrome is a condition in which extra genetic material causes delays in the way a child develops, both physically and mentally. When a child has Down syndrome or Pierre Robin sequence, he or she can develop obstructive sleep apnea due to delays in development and anatomical changes of the tongue and jaw complex.

Children with Down syndrome are at an increased risk for sleep apnea and should be closely watched for the development of these symptoms.

- weight loss or poor weight gain
- mouth breathing
- enlarged tonsils and adenoids
- problems sleeping
- excessive daytime sleepiness
- daytime cognitive and behavioral problems

Obstructive sleep apnea occurs most often in children with Pierre Robin syndrome due to an airway obstruction, small jaw, and other conditions. Through this, children have difficulty breathing and sleeping at night.

Through treatment, children can learn to breathe and feed properly while optimizing growth and nutrition, despite the predisposition for breathing difficulties. If there is evidence of airway obstruction, the child's sleeping position will be altered to help bring the tongue base forward. With proper treatment, children with Down syndrome or Pierre Robin sequence will be able to sleep better at night.

Hypothyroidism

Obstructive sleep apnea and hypothyroidism are relatively common disorders with similar clinical features.

Hypothyroidism is a condition in which the body produces insufficient amounts of thyroid hormones, affecting many complications in the body including fatigue, loss of muscle tone, weight gain, and sleep apnea. Hypothyroidism and sleep apnea are linked because hypothyroidism leads to swelling of tissues everywhere. This includes the tissues in and around the voice box and throat. The tongue is also enlarged.

Thyroid hormones are necessary for normal growth, muscle development, and basic cellular metabolism. Sleep apnea is a condition in which the body constantly awakens a person because breathing is stopped while sleeping. The link between sleep apnea and hypothyroidism is due to the more frequent symptoms of hypothyroidism, which are swelling of the tongue and other tissues that line the mouth and throat.

In a prone position, it is easy for the enlarged tongue to block the airway to the lungs, especially since the other tissue surrounding the airway is already partially hindering airflow. Frequent sleeping on your back greatly increases risk, but even sleeping on your side will not prevent all instances of sleep apnea.

Conversely, there is no evidence that sleep apnea causes hypothyroidism. Therefore, we know that it is a one-way street and hypothyroidism is causal to sleep apnea. Additionally, another frequent symptom of hypothyroidism is obesity, which is one of the leading causes of sleep apnea. Because of this, treatment for sleep apnea is often not needed once a person is diagnosed with hypothyroidism, since treatment of this condition will decrease the instances of sleep apnea.

Gigantism and Acromegaly

Gigantism is a condition with extremely tall stature which results from too much growth hormone. When too much growth hormone occurs after the person has already reached their final height, only the hands, feet, and facial features are prominently enlarged. This is acromegaly, a syndrome that results when the anterior pituitary gland produces excess growth hormone (GH) after epiphyseal plate closure at puberty. This condition most commonly affects adults in the middle age range and can result in severe disfigurement, complications, conditions, and premature death if undiagnosed. Due to the enlargement of the tongue and thickening of the tissues of the larynx, many patients with acromegaly experience sleep apnea.

Sleep apnea is a common complication of acromegaly, as it has a negative impact on quality of life and survival. The cranial deformations and hypertrophy of upper airway soft tissue are responsible for the occurrence of sleep apnea in patients with acromegaly. Through successful treatment of acromegaly, patients can experience an improvement in the severity of the complication of sleep apnea. At diagnosis of acromegaly, it is important to evaluate patients for sleep apnea because treatment can only seldom reverse it after a long period of active acromegaly.

With proper diagnosis of sleep apnea at the initial development of acromegaly, treatment is more successful and easy to maintain. Treating both acromegaly and sleep apnea is extremely beneficial in maintaining optimal overall health, which is why it is important to seek diagnosis and treatment as soon as symptoms develop.

Awaken the Giant Within, a popular self-help book by motivational speaker Anthony Robbins, is a call to "wake up and take control of your life!" Since snoring and sleep apnea are common in both gigantism and acromegaly, this gives a whole new meaning to "awaken the giant within!"

Hormones and Stress

The changes that obstructive sleep apnea affects on hormones can be summed up in one word: stress. There is an old pneumonic—ABC—taught in CPR class. It stands for airway, breathing, and circulation. It sums up the amount of stress that is caused by obstructive sleep apnea, and *nothing* is more stressful than not breathing.

The brain's stress response to not breathing alters the hormones that are sent forth from the pituitary. Everyone has probably heard of cortisone. It is also known as cortisol and is regulated by a hormone put out by the brain called ACTH. Too much cortisol causes a condition known as Cushing's disease, the classic feature of which is the tendency for those affected to get fat in the in the trunk of the body. We know that sleep apnea causes alterations and the cortisol secretion process. This may be one of the reasons that people get fat when they have sleep apnea. It may also be one of the reasons that they tend to lose weight after successful treatment.

This same hormone tends to cause us to increase the amount of sugar that floats through our blood. In times of stress, it's useful to have the extra energy. However, the insulin that we make also tells us to store this energy, and it tends to be stored in fat cells. This stress response is also why we tend to become less well controlled in our sugar metabolism, resulting in a worsening of diabetes. This is one of the reasons that sleep apnea therapy tends to improve diabetes control. In fact, insulin resistance is a higher risk factor for sleep apnea than the body mass index, which is a measure of body fat.

The metabolism of leptin is also altered in patients with sleep apnea. This hormone is commonly known as the satiety hormone. It is released from fat cells to tell us to stop eating so much. It also increases respiratory drive, which is how it's connected sleep apnea. Patients suffering sleep apnea have high circulating levels of leptin, which should tell them to stop eating, but the brain seems to become insensitive to leptin in sleep apnea patients. The brain in patients with sleep apnea is also secreting ghrelin, a twenty-eight-amino-acid hunger-stimulating peptide and hormone, which also can be a key factor in weight gain.

One final observation is that patients with high blood pressure often get better after receiving treatment for sleep apnea. Under stress, the brain increases production of renin, a hormone that affects blood pressure, among other things. Doctors commonly treat the hypertension with ACE inhibitors, such as lisinopril or enalapril. Treating sleep apnea tends to lower the production of renin, and this may be one of the reasons that successful treatment for sleep apnea improves blood pressure control in many patients.

Being excited or frightened increases the release of adrenaline (epinephrine). There's nothing more frightening than being smothered. When sleep apnea occurs, catecholamines, including epinephrine, are released. It appears that the arousals precipitated by the apnea during sleep increase the release of these catecholamines. Patients who have sleep apnea have increased catecholamine blood levels. This finding may be an additional reason for the hypertension associated with sleep apnea, yet another good reason to seek treatment.

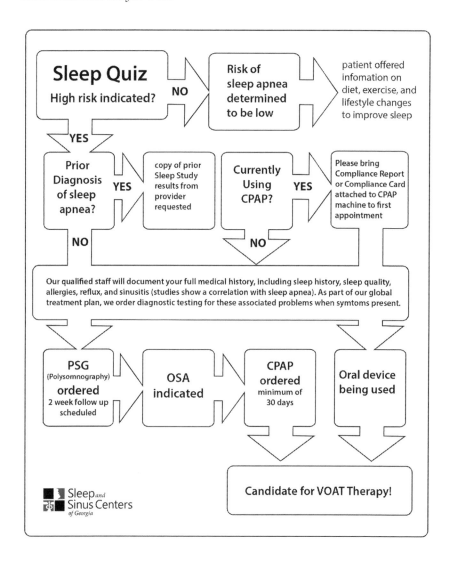

Outline of our diagnostic process to determine
appropriateness of VOAT therapy.

Chapter 5

GERD: Gastroesophageal Reflux Disease

There is nothing more deceptive than an obvious fact.

—Sir Arthur Conan Doyle, "The Boscombe Valley Mystery"

Reflux has been mentioned in past chapters as a contributing factor for sleep apnea. Let's begin our discussion by defining what reflux actually is and how it contributes to, and/or results from, sleep apnea. Most people who have sleep apnea or nasal problems, or problems with the voice box for that matter, are unaware that they have reflux. They can sometimes experience terrible reflux but not even know it. In our practice, we test for reflux in patients that have sleep apnea and are not proceeding as well as we would like. Surprisingly, it is more common than not. In my practice, almost 90 percent of patients who have severe sleep apnea seem to have moderate to severe acid reflux, with most not even knowing about it.

In the last fifteen to twenty years, it has been demonstrated that acid coming up from the stomach and, quite possibly, the volume of fluid coming up from the stomach into the esophagus and the upper airway cause a great deal of problems in the throat. It doesn't take a lot to imagine that the body tries to protect itself from having acid in the lungs. It seems that acid does a large number of things in the upper air way and has been demonstrated to have the ability to activate a protein that we use to digest things. This protein is called pepsin, which breaks down steak, meat, and other proteins in the stomach.

Pepsin has now been discovered throughout the upper airway, including inside the eardrum—the middle ear. Dentists have long known that stomach acid destroys tooth enamel. We can find evidence that stomach contents damage the nose and the voice box as well. Stomach problems can even contribute to lung problems and sleep complications.

For many doctors, the association with sleep apnea and acid reflux is a relatively new concept. However, we have known of this association for some time and now champion awareness. If you don't know what acid reflux is, it is like a little episode of vomiting into the nose and throat. The term *throat* is used loosely in this instance and includes the esophagus and all other systems in the back of the mouth.

Also known as acid reflux, GERD is an acronym that stands for gastroesophageal reflux disease. This is a chronic illness that affects about 5 to 7 percent of the world population and is often associated with serious medical complications when left untreated. When investigating GERD, it becomes apparent that this disease and sleep problems go hand in hand. When most people think of acid reflux, they often think of heartburn, which is actually the body signaling damage to the esophagus.

While researchers and physicians are still working to understand the link between GERD and sleep apnea, we are aware of one clear connection. In our experience, between 50 and 75 percent of patients with sleep apnea also have GERD. Some of the connections between these two conditions include

- interrupted sleep
- relaxation triggers
- CPAP side effects

People who have the serious symptoms of GERD report experiencing the worst quality sleep, but the nature of the sleep interruption is still not fully understood. Nighttime reflux is often painful enough to disturb sleep or wake up the person—up to 30 percent or more of our patients with GERD complain about this nighttime reflux or heartburn. However, this percentage is difficult to gauge, as many people are unable to recall waking up in the middle of the night. As such, the real percentage of patients with severe apnea who test positive for moderate to severe reflux seems to be 80 percent or more in our experience.

If you suffer from GERD, you might also be more likely to experience reflux symptoms during periods of wakefulness at night. This can happen frequently if you also have sleep apnea. As you wake up throughout the night, the lower esophageal sphincter relaxes, thus being awakened by an apnea event could trigger your reflux.

Additionally, many patients who suffer from sleep apnea undergo treatment with continuous positive airway pressure (CPAP), which can trigger side effects. The CPAP machine constantly feeds air into your system to prevent the interrupted breathing common in sleep apnea. But for a small number of patients, this treatment may expand the stomach slightly, which leads to more reflux in some sleep apnea patients. For this reason, we offer an array of other treatment options that—when compared with CPAP—uniquely eliminate your sleep apnea symptoms while also improving your GERD symptoms.

GERD Case Study

A patient with sleep apnea who came to our clinic was interested in undergoing a tongue reduction procedure. During the course of his evaluation, he was tested for acid reflux. He insisted that he did not have any reflux, but after a lot of coaxing—much of it coming from his wife who had resorted to sleeping in another bedroom—he was convinced to go see a surgeon to reevaluate his reflux. This patient was a big, healthy, and athletic-looking guy who was arguing in the most polite terms with his wife that he did not need to have anything done. But she knew better! One of the reasons that married men live longer than single men is that wives have some kind of supernatural method of making them seek medical attention, even when they don't want to. This patient's wife was no exception. He ultimately wound up having the necessary reflux surgery.

Without further treatment, this patient's sleep apnea score went from 35 to less than 5. Effectively, he was cured of his ailments. Treatment of sleep apnea and GERD appears to help patients in the reduction of their symptoms. If you improve acid reflux, you also improve sleep apnea and vice versa. However, it is not known yet whether one condition causes the other. The key to treatment is to let your doctor know that you have both conditions. By understanding this, you can undergo treatment to improve both sleep apnea and acid reflux.

By seeking treatment for your sleep apnea, you can further decrease incidence of GERD, just as treating GERD can also decrease the symptoms of sleep apnea. If you are obese, reducing body weight can also decrease the symptoms of both GERD and sleep apnea. Nocturnal GERD and sleep apnea can lead to serious complications and quality of life issues. Taking the next step by treating sleep apnea and GERD will help you maintain a healthy, active lifestyle without complications. Act now, and protect your well-being.

GERD Case Study #2

Recently, a patient experiencing symptoms of sleep apnea came into our office. He did not present as overweight, or unhealthy in appearance. On the contrary, he was a man in his midforties who took good care of himself. He watched his diet and worked out regularly. In short, he did not fit the profile.

His complaint was not atypical. Every night, he faced the all too common scenario of his wife poking him in the ribs, desperately trying to get him to roll over and stop snoring. Both his life and his wife were dramatically impacted by the all-too-apparent *symptoms* of sleep apnea. His thunderous snoring and gasping and choking sounds occurring every night led neither of them to feel like they had gotten a decent night's sleep, even after sleeping for many hours. They both experienced continuous feelings of fatigue, exhausted mornings, and daytime sleepiness. The patient's wife was also kept awake at night worrying that her husband would stop breathing long enough to cause serious harm. The couple's continuous lack of sound sleep led to sluggishness and a decreased ability to focus on everyday personal and professional tasks. It negatively impacted conversations, caused mood swings, and eventually resulted in depression.

All aspects of their life were negatively impacted, and they felt powerless to change their situation. What could they do? Remember the patient *seemed* to be living a healthy lifestyle.

Being patient centered in our approach, we were determined to diagnose and treat by looking at all facets of his situation. We ran a comprehensive spectrum of tests that would lead to a more thorough analysis of his distinct condition. Testing showed us that his voice box was swollen. Further analysis led us to a diagnosis of severe GERD. It is important to note that the diagnosis in this case was not only sleep apnea, but also sleep apnea and laryngeal edema (that is voice box swelling) and sever reflux. His symptoms might easily have yielded a premature diagnosis of sleep apnea only. In defense of the potential missed diagnosis, the CPAP that presumably would have been ordered to treat the sleep apnea might have mediated some of the patient's symptoms. However, it would not have addressed the root cause. Recognizing the reflux he was suffering as an important clue, we immediately had his reflux treated with a transoral incisionless fundoplication (TIF), an incisionless, endoscopic procedure.

As the patient started to experience relief from his reflux, his mutually exhausting, disruptive sleep apnea episodes started to disappear as well! Post-TIF, this patient's breathing was absolutely normal. Without further treatment, his sleep apnea score went from 40 to essentially normal. Effectively, he was cured of his ailments. Treatment of sleep apnea and GERD appears to help patients in the reduction of their symptoms. If you improve acid reflux, you also improve sleep apnea and vice versa. However, it is not known yet whether one condition causes the other. The key to treatment is to let your doctor know that you have both conditions. By understanding this, you can undergo treatment to improve both sleep apnea and acid reflux.

By seeking treatment for your sleep apnea, you can further decrease incidence of GERD, just as treating GERD can decrease the symptoms of sleep apnea. If you are obese, reducing body weight can also decrease the symptoms of both GERD and sleep apnea. Nocturnal GERD and sleep apnea can lead to serious complications and quality of life issues. Taking the next step by treating sleep apnea and GERD will help you maintain a healthy, active lifestyle without complications.

His wife is now greatly relieved to report that she no longer wakes up multiple times every night to the sound of his snores or gasps, nor does she feel the need to check if he is breathing multiple times throughout the night. They both have experienced increased feelings of health and energy and feel more capable of concentrating on their important daily activities. The patient no longer feels like his emotions are unceasingly controlled by his lack of rest, and his quality of life has greatly improved.

This patient's story illustrates a few important points. First, that eradicating causal sources of sleep apnea can largely eliminate the effects of sleep apnea and improve one's quality of life; second, that reflux, and more specifically GERD, can contribute to the life-disrupting trials caused by sleep apnea; and third, that reflux cannot always be identified by heartburn. This patient did not come in to our office complaining of heartburn. He did, however, report that he had always had some mild nasal problems. He sometimes had trouble swallowing and had dry spells intermittently. These issues are symptoms that one does not usually think of as being major associations with reflux.

The big-picture view of how reflux relates to sleep apnea is this: when a person affected by reflux is sleeping, he will unconsciously clear his throat, which brings up and spreads around the acid afflicting him. The acid will irritate him and cause harmful chemical changes within his body, which can lead to damaged lungs, upper airways, and digestive tract, as well as causing swelling in his throat and nose. The swelling eventually creates pressure, which in turn increased the likelihood that he will experience reflux. This swelling, as well as the tongue's movement to clear the throat, prevents the easy flow of air to the lungs, contributing to episodes of sleep apnea. Victims of sleep apnea also tend to breathe harder because their breathing has stopped, which could induce reflux to flow into the esophagus. In this way, reflux and sleep apnea perpetuate each other; the results over time can be extreme and can prepare the body to become more susceptible to more dangerous diseases.

The vicious cycle of reflux and the resulting episodes of sleep apnea will continue more and more frequently and intensely unless the GERD is addressed. Even if a treatment for sleep apnea is prescribed, episodes can persist if its underlying cause is not eradicated.

A common misconception for sleep apnea patients, as well as for the vast majority of people, is that reflux is a burning sensation. Patients will come in complaining of a chronic tickle, something that feels like a frog in their throat, or stuff they can never cough up. Or they may have sinus problems. What they don't realize is that all of those symptoms are indications that they could be suffering from reflux.

The effects of reflux can manifest in a myriad of ways. This is how it happens: Reflux is similar to vomiting in that people who are vomiting are affected by burning acid and have trouble breathing during the event. Comparatively speaking, reflux is just like little bits of recurring vomit. The small problems that you get from vomiting are not too bad if the vomiting doesn't last too long, but if you're chronically exposed to it, your body has to try to do something to fix it.

The body may cough, a reflex designed to clear its airways, or may produce thicker saliva to help calm down the agitation. Many times, what these patients are actually experiencing when they complain of a chronic tickle or something they can't cough up are actually small amounts of a very thick, clear mucus or even a thickened voice box or esophagus.

To understand these symptoms of reflux and how they relate to sleep apnea, it is important to understand how the esophagus is related to the stomach, voice box, and upper airways and how reflux affects all of these major components of the digestive and respiratory systems.

The esophagus is the first part of the body affected by reflux. Like a football player, the esophagus is tough, having learned through years of training how to take a punishing beating.

The upper airways, on the other hand, are relatively weak. They are like someone who has never exercised (or to continue with the metaphor, never held a football!). They also have very little padding to protect them from stomach acids. In short, the upper airways are too weak and ill equipped to handle the "offensive surge" of these acids.

The esophagus's "pads" in this analogy are its mechanism for neutralizing the acid from the stomach, which is called bicarbonate. The mechanism is known as carbonic anhydrase (CA), and the esophagus has a ton of CA, as well as many mucus-secreting glands, which provide a tougher type of lining than the material in the upper airways, devoid of any neutralizing CA. The voice box and pharynx (upper throat) relies largely on saliva to solve the problem acid creates on entry.

The throat also has many tiny salivary glands scattered throughout that help deal with acid from reflux. The glands produce mucus and saliva to neutralize the acid. They are the part of our body that is responsible for getting moisture into the mouth and for helping to protect the throat from damage.

While the esophagus is better able to protect itself from the effects of acid than the upper airways are, it is not immune to them. To better understand how acid affects the esophagus, imagine sitting at a table with a lemon and squeezing it. If the acidic juice reaches your eye, your eye gets red and irritated. It hurts! The same sort of reaction goes on when you have acid coming from your stomach into your esophagus. The acid irritates and inflames the esophagus. In fact, when reflux is recurring, it chronically makes the esophagus swell, triggering and activating all kinds of growth factors that are local hormones. Reflux may lead to esophagus ulcers and possibly cancer.

The apnea reflux might be thought of as the body trying to protect itself from acid reflux, because it doesn't want acid in the important airways, including the lungs. If acid bothers your eyes, you know it's going to bother your lungs. Untreated reflux can lead to decreased lung capacity and other forms of respiratory damage. The body will react in as many ways as it can to protect the lungs because they are so vital to life, so fragile, and so irreplaceable.

The nose is one of the most commonly affected upper airways. It can be thought of as a bony box. When swelling occurs inside the nose, there is no place for the swelling to go but inward. This makes the nose feel like a clogged drain that nothing can move in or out of.

It can't empty mucus very well, making breathing difficult. Because most people sleep on a flat surface, mucus drainage can be extremely slow or even impossible at night, contributing to episodes of sleep apnea. The swelling and mucus can also lead to the development of rhinitis and other conditions impacting systems that perform important breathing functions and noticeably impact everyday life.

Similar occurrences of swelling and possible infection can happen in the ear due to reflux. The ear is connected to the back of the nose, which is connected to the network of upper airways, which is connected to the esophagus, which is connected to the stomach. This is evidenced by the fact that a protein called pepsin, an enzyme in the stomach, has been found as far away from the stomach as the ear! Clearly, reflux is far-reaching and potentially dangerous, particularly when it contributes to the effects of sleep apnea.

Pepsin does many things that contribute to reflux, and by extension, sleep apnea. It can make the lining of the voice box become swollen and sore, and that alone may cause problems.

The thickness alone can make you feel like you've got drainage in the back of the throat constantly, because the lining feels foreign, not a part of you. To understand why this happens, consider a "fat lip". When you suffer a blow to the lip, the flesh on the lip swells and the nerves on the flesh spread out, causing the lip to feel foreign, like someone else's lip attached to your face.

The same process and feelings that accompany a "fat lip" happen to the lining inside of you when pepsin causes it to thicken. The thickness makes it feel like something is clinging on to the throat, but that "something" you can't get rid of because it is part of your body!

When the voice box is affected by pepsin and reflux, the symptoms can manifest in several audible ways. The voice tends to get hoarse and the tissues that lie over the windpipe tend to get floppy and can make a noise when they get sucked into the windpipe. The resulting sound is somewhat like asthma, for which it is often mistaken. Reflux can cause spasms in the voice box as well. Finally, the inflammation and irritation can reposition the covering to the voice box, and this can permanently cause it to be prone to obstruction like in sleep apnea.

Another problem caused by pepsin's appearance in places other than the stomach is that it can manifest itself in the form of a chronic cough, which comes from the thickening of the lining around the voice box. The cough reflex is designed to clear material from out of the voice box. When the voice box is swollen, it shakes like a bowl of jelly. It vibrates and tickles and triggers one to protect the airway through the cough reflex. It can also trigger involuntary breath-holding events, and therefore sleep apnea.

Additionally, because one needs saliva to neutralize the acid in the airways, which are relatively unprotected areas, the effects of pepsin in them can cause one to develop a water brash, which is the regurgitation of an excessive accumulation of saliva from the lower part of the esophagus, often combined with some acid material from the stomach.

The body will strive to remove the excess saliva and acid to protect the airways and their paths by swallowing. However, one of the reflexes that allow one to swallow also stops one from breathing. In this way, the sleep apnea episodes can occur.

The tongue's response to reflux can also play a major role in causing sleep apnea. The tongue is designed to protect the voice box from swallowing and from reflux. It does so by closing the epiglottis, which is like a toilet seat flap that flips down over the opening of the trachea, creating a seal that does not permit anything other than air to enter it. This anatomical structure is very important, as without it an organism would run the risk of choking and coughing every time it tried to eat. However, when affected and irritated due to reflux, it can also cause you to hold your breath.

The tongue can also be unknowingly exercised all night in an effort to clear the airways around the larynx and around the nose; this incessant nocturnal activity may be one of the reasons it gets larger and therefore big enough to block the airway and stop one's breathing. It can also collapse backward during sleep as the muscle relaxes, particularly when one sleeps on his back, ultimately leading to episodes of sleep apnea.

The link between reflux and sleep apnea is significant and dangerous. However, each problem can be treated and cured, and your life can return to its normal, healthy state.

A few simple lifestyle changes may help you combat reflux.

One of those changes is to stay elevated during sleep. Because water runs downhill, you will want to try to keep your head elevated above the acid in your stomach, so that the acid does not run toward it. In this way, you can help prevent it from easily flowing into your esophagus and upper airways. Keeping your head at least four to five inches above your stomach is a good goal when trying to stay elevated in a therapeutic manner.

In addition to staying elevated during sleep, there are several dietary changes that can help reduce the effects of reflux.

One easy change you can make is to not have any food in your stomach when you go to bed. This means not eating a full four to five hours before you go to sleep. In order to get as much energy out of a piece of meat as possible, the body wants to keep it in the stomach, because the stomach's acid environment—in conjunction with that protein—chews the food up most efficiently, and therefore gets the most use out of it. Not only does not eating before bed help you get the most benefit from your diet, it also can reduce episodes of reflux. Not eating close to bedtime also helps you protect your airways by helping them stay clear of any clogging debris and obstructions.

Another dietary change you should make goes along with the first rule. That change is to eat your largest meal in the morning and eat progressively smaller meals as the day goes on. An old saying goes "Eat like a king in the morning, a prince at noon, and a pauper at night." This old maxim is full of wisdom, which is probably why it has stuck around. In fact, our friend the dirt farmer from chapter 1 most likely subscribed to this healthy habit. Eating in this way helps you keep your stomach empty before going to bed, which helps your body absorb as much energy as possible from the food you consume, with the added benefit of keeping your airways clear!

A third and final dietary change that will help mitigate problems with chronic reflux comes in the form of another old adage: everything in moderation. Specifically, you should consume spicy foods, tomato-based foods, acidic foods, chocolate, and high fatty meals in moderation. Additionally, in general, you should only drink water. While it is true that most beverages, such as carbonated drinks, alcoholic drinks, and caffeinated drinks *are* mostly made of water, if they contain anything other than water, and you are serious about curing your reflux, you should probably not consume them.

The reason these foods should be avoided for the most part is that the body is not designed to handle a large influx of them on a continuous basis. In the past, many of these foods were consumed only on special occasions. Only recently have they been easily and readily available to humans, and our bodies have not yet been prepared to handle them without negative repercussions.

While dietary choices can contribute to reflux, they are not its only cause. Another cause of chronic reflux could be hiatal hernias.

To understand hiatal hernias, you should first know that there are natural mechanisms that keep the acid from coming up from our stomach all the time. They are the upper esophageal sphincter and the lower esophageal sphincter.

Hiatal hernias can be understood as a stretched-out sweater: the spring has been taken out of it, and it no longer fits right. In the case of hiatal hernias, the stretched-out opening of the sweater is analogous to the opening to the underside of the diaphragm. When that is loose, all of contents of the stomach are able to go through the lower esophageal sphincter a lot easier, allowing acid to flow into places it was not designed to flow to, where it then can cause damage.

If you have acid reflux, you may need to put your sweater in the dryer—that is to say, have your lower esophageal sphincter tightened!

While there are many symptoms of reflux, the only way to know if you have reflux for sure is to have somebody look at your airway, voice box, and inside the nose to look for signs that you have reflux. Sometimes, it can't be seen, but it can be tested.

In evaluating your own ordeal with sleep apnea, there are a few things you should consider if you think you might also have reflux.

It turns out that 90 percent of patients I see who have significant reflux also have sleep apnea. Reflux is worse when one is lying down, as in sleep. At least 80 percent of the sixty million Americans who have been diagnosed with GERD report worse symptoms at night. Three in four say they routinely wake up from sleep because of them. If you have heartburn more than twice a week, you may have GERD. But you can have GERD without having heartburn.

Reflux can cause a chronic nasal and throat irritation from the voice box up, which often goes underdiagnosed. Reflux can also be extremely dangerous to any affected person because his unconscious actions during sleep help perpetuate the reflux, as well as contribute to the occurrence of episodes of sleep apnea.

Reflux is not always recognized for what it is, and many patients do not know that it can be a source of sleep apnea. Reflux is also often underdiagnosed; when it is diagnosed, its severity is not always recognized or addressed.

This problem arises because the methods many doctors use to explore a patient's reflux look for the damage it causes, but not the process by which it arises or the way it could be affecting the patient. This inefficiency occurs because the typical methods used, such as a barium-swallow radiograph or an endoscopy, are not capable of identifying many of the episodes of reflux that may be afflicting a patient.

Because reflux often occurs at night, and because the reflux has to occur during a very short window of time, such as when the radiologist is checking for it, it is often missed during a doctor's evaluation.

To understand why these methods are inefficient, imagine a cheery fellow who always has a smile on his face and a kind word to share. He has just told a joke that made those around him laugh, when suddenly, he steps on a nail! His face screws up into an expression of anguish, and at that moment, someone snaps a photo of him and the group he has just made laugh.

Later, someone who does not know the man finds the picture. In it, he sees a group of people looking happy and good-humored and a man standing next to them looking very upset. He thinks, *That man must have a bad-humored personality to be so unhappy in the company of such a joyful crowd.*

Of course, this assessment is inaccurate and unfair, because the man examining the picture does not know anything about the circumstances that created it. In the same way, viewing a patient's esophagus at a specific moment in time does not give someone enough information to diagnose its overall condition.

Another way to appreciate why these methods cannot fully diagnose the severity of a patient's reflux is to compare them to a doctor only looking at the surface of a patient's skin when attempting to diagnose a case of skin cancer. Even if a doctor could tell that a patient had skin cancer this way, he would not be able to identify anything about its origins or its full range of affliction. Similarly, a doctor cannot identify where the path acid from the reflux is coming up from one image of it at a single point in time.

A more efficient method of understanding and testing the severity of GERD is through a twenty-four-hour pH test. This test is an outpatient procedure that measures the pH or amount of acid that flows into the esophagus from the stomach during a twenty-four-hour period. It is taken by using a small probe that is inserted through the nostril and positioned near the lower esophagus. The probe is plugged into a small monitor that stays on a belt or over the shoulder. A newer, wireless device may make monitoring the pH level easier, allowing a doctor to place a disposable capsule into the esophagus using an endoscope. The capsule then wirelessly transmits information to a receiver worn around the waist. This newer method is more comfortable than the previous.

This test gives a more complete picture of what the patient is experiencing in regards to reflux episodes and can therefore lead to more accurate diagnoses of the reflux's severity.

Accurately measuring the severity of a case of reflux is important because when the reflux is underdiagnosed, a patient will often be put on some form of medicine. Zantac, Prilosec, Tagamet, Aciphex, and their generic forms are medications that are often prescribed for patients suffering from reflux.

The problem with simply addressing reflux by taking a pill is that the treatment does not necessarily keep the fluid from going up into the airway. Whether the acid is neutralized by the medicine before it gets into the lungs is rather unimportant; regardless of whether orange juice or water was to get into the lungs, it's a problem.

Additionally, the medicine itself may cause harmful side effects. Elderly individuals with thinner bones, in particular older women who are affected by osteoporosis, are impacted most. Unfortunately, medicines typically prescribed for stomach problems may actually aid in washing out their bones.

Other issues that these medications can cause include gastric polyps, which are masses of cells that form on the inside lining of the stomach which may increase one's risk of developing stomach cancer. The medications can also potentially increase the risk for pneumonia. If the reflux is not completely eradicated, a patient may be at increased risk of developing esophageal cancer.

We frequently find that patients suffering from reflux will visit the doctor for their annual visit and will simply be put on medication to address their symptoms. While taking medicine is better than not doing anything, their condition could be greatly improved through procedural interventions.

In the past, there has been a significant failure rate with surgical interventions. However, in recent years, procedures such as the TIF (transoral incisionless fundoplication) and a variety of other minimally invasive procedures have improved greatly, and if dismissed previously, now deserve reconsideration.

While the idea of many of these procedures might scare you, consider how Franklin Delano Roosevelt inspired a frightened nation suffering in the grip of the Great Depression with the strength of his oratory on the occasion of his first inaugural address. He said, "The only thing we have to fear is fear itself."

As was true then is still inherently true today: not doing anything is always worse than addressing the problem head on. If medications have not effectively eradicated your problems with reflux in the past, look into TIF and other new procedural options. What may have seemed intimidating in the past might be an important first step in your path toward real and lasting—pardon the political pun—"recovery."

Chapter 6

Sinusitis

A body can get used to anything, even to
being hanged, as the Irishman said.

—Lucy Maud Montgomery, *Anne of Green Gables*

The nose is a critical piece of the sleep apnea puzzle. It is an important breathing mechanism, because our bodies are not designed to breathe through the mouth if at all avoidable. Nasal resistance is a major source of problems that lead to sleep apnea.

A common problem which we have observed is that patients with sinus problems intrinsic to their respective cases of sleep apnea often do not make the connection between the two. Not only do these patients fail to realize that their sinus problems contribute mightily to their sleep apnea, but incredibly, they frequently don't acknowledge that they have any sinus problems to begin with!

This common yet curious finding is best illustrated through a narrative case study of a real patient. The patient was a man who had been suffering sleep apnea for some time; his symptoms included heavy snoring and trouble breathing. This patient had already taken many steps to treat his sleep apnea. Most notably, he had been put on CPAP. However, the machine did not work very well, and he very frequently felt the need to remove it. In an effort to further deal with his problem, the patient tried all of the usual measures: sedation, sleeping aids, different masks, and even a dental device. Nothing seemed to give him the relief he sought, and needed.

The patient did not believe that sinus problems could be causing his failure with the CPAP machine; he did not believe that sinus problems affected him. After all, he felt no facial pressure, and he experienced no drainage. None of the typical symptoms of sinus problems were present. However, after taking his history and speaking with him for a while, we found that he had always had a diminished sense of smell.

We took a look at his nose and found that he had a giant polyp growing inside of it. It occupied all of the space on one side of his nose, leaving it 100 percent blocked, and it also took up half of the space on the other side. Material from the polyp filled up so much of his nasal cavity and sinuses that there was very little room for airflow; yet the patient thought he was breathing fine and did not recognize that he had any sort of sinus problem.

We took care of his sinus problems by operating and removing the polyp. His sleep apnea improved so dramatically after the nasal surgery alone that he hardly felt the need for his mask anymore.

The takeaway from this patient's story is first, that sinus problems can contribute to episodes of sleep apnea, and second, that patients suffering from sinus problems may not be able to recognize that they have a problem at all. This patient's polyp had developed so gradually that he couldn't smell or breathe, and he didn't even realize how abnormal that was because he had become so used to his condition. Just like this patient, you can get used to anything if it affects you long enough. It's a scary but poignant truth when it comes to the sinuses. Problems can develop so gradually that you may not be able to recognize that you're not even breathing.

This patient's case inspired us to take an informal study in the office. We looked at many patients who, like the patient who inspired the study, were not able to tolerate their CPAP mask. We found that about one third of them were not just affected by conditions like claustrophobia but also had some sort of problem show up in the CT scans of their sinuses. Treating the sinus problems that were "exposed" helped many of these patients breathe much more easily—enabling some to even put away their CPAP!

A lack of understanding about how the sinuses work makes it difficult for many people to comprehend the tremendous impact the sinuses have on sleep apnea. The sinuses can be thought of as two hallways positioned adjacent to each other, separated by a common wall. Continuing the architectural analogy, there are steam rooms at the end of each hallway, the other side left open. They are highly humid, hollow spaces that store air briefly in an environment with a high moisture content. Their function is to provide constant humidity to the air streaming into and through them. That mixing of the air with warm humidity allows them to decrease the weight of the air and keep airflow light and easy.

To understand how the sinuses decrease the weight of the air, let's compare them to another system: a weather pattern. Meteorologists commonly refer to low-pressure and high-pressure zones; the low-pressure zones come out of warm places like the Caribbean or the ocean. The hot, wet climate of the ocean warms up the air, creating low-pressure zones that flow toward high-pressure zones comprised of cold, dry air. The warm air weighs less than the cold air, creating the dynamic motion and transfer of energy defined by meteorologists as a weather system.

The sinuses are like a weather system on a smaller scale, all contained within your body. The principles work the same way: properly functioning, healthy sinuses warm the air flowing through them, and that allows air to pass easily and with low pressure.

The sinus also works as a filter. It is a mechanism that allows a constant flow of air to new mucus. The warm air helps the mucus work its way out. If the holes in the sinus become closed off, it cannot create the warming effect and can become swollen. Oftentimes, a patient will come in who has lost his sense of smell. The loss of olfactory sensation most often has occurred because the sinuses have become so swollen that particles can't come in through the nose into the area where the smelling occurs. The loss of smell is a fairly reliable indicator of sinus problems requiring medical intervention.

Another common issue that patients face is that they start to feel like their nose is abnormally dry. Sometimes, the nose gets so dry that it bleeds frequently. This happens because the moisture is being extracted out of their mucus, moisture that the sinuses are unable to replenish when inflamed or otherwise blocked.

The CPAP mask contributes to and can often significantly exacerbate sinus problems. To review the function of a CPAP machine, imagine a pipe with water running through it. How do you get more water to flow through it? You can either make the pipe bigger or you can increase the pressure of the water running through it. The CPAP machine works by increasing the pressure of the air flowing through your nose, in an effort to deliver more of it to your body.

When air is pumped through the mask, it tends to dry out the sinuses. Moreover, because that air is relatively compressed, it sucks the moisture out even further. You can think of it like winter air flowing from the cold outside into a hot house; there is not much moisture in the air in the winter to begin with, and when you warm it up, it becomes even drier. That invading, cold draft is analogous to what happens when the sinus is not able to moisturize the air going through it. The "untreated" dry air is not very good for the lungs, and the lungs are not able to work with it very well.

It is common for patients to develop nosebleeds when they begin using the CPAP machine. Compounding this tendency, some of these patients may be taking blood thinners, may already be sick, may have really high blood pressure, or may have other medical conditions that facilitate bleeding. However, one thing that is fairly common among them is some sort of sinus problem, whether it exists in the form of allergies, rhinitis, or other forms of autoimmune nasal problems. These patients are frequently unable to keep the masks on.

The nose is like a finely tuned machine that needs to be working in perfect harmony in order to function effectively. With every part working in concert, we breathe well, moisturizing the air we take into our lungs, effortlessly. Dryness, bleeding, and uncomfortable pressure are symptoms that are easy to recognize. What's more subtle and sometimes hard to pinpoint is when some small portion of the nose is out of whack and you're not doing any of those things well.

Imagine that your nose is an exotic car, perhaps a Lamborghini. If your Lamborghini were out of tune, maybe it would not sound quite right to you. Respectful of the precision engineering and complexity of this world-class head turner and pulse enhancer, you would not simply get under the hood and tinker around. You would not dream of doing that! You would take it to a professional, factory-trained, and experienced mechanic. Moreover, the mechanic's shop would be equipped with the latest computerized diagnostic equipment available to most effectively determine what was robbing your baby of its peak performance.

In the same way, you may not have the answers for what is going one with your nose, although you are trying to deal with it. Certainly, any fool can look at a Lamborghini, see that the tires are flat, and put more air in the tires. In the same way, there are some relatively simple, almost universal actions that we can take to significantly improve the condition of the nose. Anecdotally, my nose is more like a DeSoto than a Lamborghini, albeit I still choose to have it serviced by a professional!

1960 DeSoto Adventurer

In the past, nasal surgery has been intimidating, causing many people to shy away from dealing with the sinuses when they have problems. However, today, in the same way that angioplasty has made heart surgery almost routine for a cardiologist, procedures like sinus balloon dilation allow greater numbers of patients find relief. In the past, fear might have prevented a patient from electing the same kind of surgery that perhaps a grandparent had once endured. Now, with the experience and technology afforded modern doctors, there is no reason to ignore a nasal problem. One thing is certain: the problem will only get worse. Fear, again, is the enemy. See your doctor. Educate yourself. You might be surprised to find that the solutions are routine and not very scary after all.

There are many problems that should be looked at in diagnosing a sinus problem. For instance, many adults still have their adenoids, which are kind of like tonsils inside of the nose. Just like stuffing a piece of meat inside a drain would clog it, adenoids can sometimes plug the back end of the sinuses, causing harmful blockages. In another case, the nose might collapse either from congenital weakness or something as benign as the aging process. As we age, our ligaments sometimes become less capable of holding the cartilage—the springy portion of the nose—out to the side of the face. Still another condition could occur in which turbinates, which are cigar-shaped structures inside the nose that expand and contract like water balloons, become filled with blood, permanently inflamed, and enlarged. Or a nasal valve could collapse. Additional conditions could include anatomic variances like concha bullosa, septal deviation, and paradoxically curved middle turbinates.

When a nasal valve collapses, it is akin to somebody pinching your nose when you breathe in. The nasal valve is the area of the soft part of the nose that's on the outside the face where it meets the bone. Imagine that there is a flag attached to the antenna on your car. If the fabric is too loose, the fabric of the flag shreds at high speeds. The flag is destroyed because the fabric is not stiff at all and is completely at the mercy of, in this case, the motion-induced wind. On the other hand, when the fabric of the flag is stiff, you can drive your car at the fastest speed you like with the flag fluttering behind because the fabric is tough enough to handle the breeze. Of course, you would never fly a flag from your Lamborghini, but that's another story …

The tip of the nose is akin to this flag on the car; the breath you breathe is the breeze caused by the speed of the car. When you sneeze or breathe out really hard, the air flowing through your nose can reach up to one hundred miles an hour, which—completing the analogy to my distinctive nose—was ironically also the top speed of the 1959 DeSoto Firedome. (Your Lamborghini will do 236 mph, which is nothing to sneeze at!) In either case, if the tip of the nose is not very stiff, it will not remain well attached to the bone and will result in collapse. After a nasal valve collapses, the material in your nose shifts and morphs, causing effective breathing to become difficult. Unfortunately, OEM parts are not available for making repairs, necessitating "modifications." While seldom medically necessary, some patients choose to have additional customizations done while repairing structural damage to the nasal valves with results that are arguably more attractive than the original!

Septal deviation is also a common nasal condition that can lead to sleep apnea. Again, using the hallway analogy, you can imagine the septum as two corridors, separated by a door. One end of the hallways is designed to be shut off at times to allow the two hallways to catch up on the humidifying process that I described earlier. Sometimes, the septum can become wrinkled; a certain degree of wrinkling, or crookedness can actually be advantageous, adding a certain amount of needed resistance and assisting mixing of air. Too much septal "crookedness" or deviation can slow airflow too much. Operations such as the one called septoplasty, can streamline the nose, just as the rivets under the wings of some airplanes effectively reduce drag and result in higher air speeds.

Too much crookedness is a problem in more ways than one. A crooked path can create tightness through the septum and not enough air will be able to get through, causing a decrease in the effectiveness of your breathing. When this occurs, the mouth opens to compensate, providing another path for the air to reach the lungs. Unfortunately, in sleep, the tongue is a like a loose hammock. When the mouth opens, the "hammock" falls to the floor, which in this case is the back of the throat. This action can stop airflow completely and lead to dangerous episodes of sleep apnea.

Sinusitis is itself a common nasal issue. Sinusitis is an inflammation of the sinus lining, which in turn causes blocked sinus passageways. Sinusitis symptoms include facial pain, tenderness and swelling around the eyes, cheeks, nose and forehead, sinus pressure or congestion, and difficulty breathing through the nose. It can present without symptoms of sleep apnea and, if chronic, should be treated early to avoid breathing problems, including OSA, in the future.

Chapter 7

Why Treat Sleep Apnea?

*If you do not change direction, you may
end up where you are heading.*

—Lao Tzu

Just like high blood pressure and diabetes, sleep apnea is a chronic disease that can impact quality of life and damage general health. We know that patients with untreated sleep apnea suffer an increased risk of cardiovascular disease, and that risk increases over time. Through treatment of sleep apnea, improvements in quality of life can be achieved while maintaining or even restoring optimal overall health. Don't ignore your symptoms. Proactive treatment can prevent further complications from sleep apnea, including conditions that could become life threatening. Let's take a look at several factors that underscore why treatment is so vital to your health.

Depression

Sleep is essential, and achieving healthy sleep should be just as important as good nutrition, physical activity, and smoking cessation in promoting overall health. Depression and sleep apnea are often linked, a finding which fortifies our understanding of just how important treatment of sleep apnea really is.

If you feel sad every now and then, it is a fundamental part of the human experience, especially during difficult or trying times. However, a persistent feeling of sadness, anxiety, hopelessness, or disinterest in things that you once enjoyed is a symptoms of depression, an illness that affects at least twenty million Americans. Not something that can be ignored, depression is a mental illness and simply will not go away with time.

Mental-health professionals often ask patients who are suffering from depression about their sleeping habits. The relationship between sleep and depression is complex yet undeniable. Studies have shown that depression may cause sleep problems and sleep problems may cause or contribute to depression. For some sufferers, depression symptoms occur before the onset of sleep complications; for others, sleep problems appear first.

According to the National Sleep Foundation and a study of 18,980 subjects in Europe by Stanford researcher Maurice Ohayon, MD, PhD, people with depression were found to be five times more likely to suffer from sleep apnea. While this connection is clear, there is good news. By treating sleep apnea, depression has been shown to improve *in proportion* to the sleep problems being diminished!

Treatment for depression involves a combination of psychotherapy and/or medications. These therapies are used to help treat both depression and insomnia, as well as sleep problems as part of an integrated depression therapy. However, treatment of depression might become complicated by sleep apnea. For instance, certain medications should be avoided when treating sleep apnea and depression due to their potential to suppress breathing, thus worsening sleep apnea symptoms. Before beginning treatment for depression and sleep apnea, talk to your physician about any sleep apnea symptoms you might be experiencing.

By effectively treating your sleep apnea, it may be enough to alleviate your symptoms of depression. Treating both depression and sleep apnea is essential in getting your life back on track. Again, doing nothing is not an option. There is too much at stake. As ancient philosopher and the father of Taoism Lao Txu is said to have observed, "If you do not change direction, you may end up where you are heading."

Heart Disease and Stroke

We all know that snoring can be quite annoying, especially if you are the one listening to it! However, when a person who snores repeatedly stops breathing for brief moments, it can lead to cardiovascular problems that can be life threatening. Sleep apnea causes pauses in breathing five to thirty times in just one hour during sleep. As stated in other chapters, these periods of pauses in breathing cause the person sleeping to wake up and gasp for air. Through these episodes, that person does not experience restful sleep and, moreover, is placed under constant stress. Risk factors increase for the patient to suffer high blood pressure, arrhythmia, stroke, and heart disease.

Heart disease is the leading cause of death in America, while stroke takes fourth place for the cause of death and is a leading cause of disability. High blood pressure is a major risk factor for both conditions. The relationship among sleep apnea, hypertension, and cardiovascular disease is very strong, which makes it vital for everyone to understand this connection.

Repetitive disruption of sleep throughout the night puts the body in a condition of severe stress, occurring with each episode of apnea. Over time, numerous adverse health consequences may result as a direct result of this stress. Cardiovascular disease, which includes hypertension, coronary artery disease, and stroke, is one of the most high-profile medical conditions today. Its impact is broad;

heart disease affects a large percentage of the population and drives enormous costs for evaluation and treatment. Cardiovascular disease is also a major cause of suffering and death, and sleep apnea is being recognized more and more as an important, and treatable risk factor for the development of heart disease.

When oxygen levels drop, carbon dioxide levels increase. As a result, the brain, sensing trouble, signals the body to release adrenaline-like substances into the bloodstream—a type of fight or flight reflex. When this occurs, blood pressure increases, thus linking sleep apnea to high blood pressure. Low levels of oxygen trigger the body to release other substances that can eventually damage the lining of blood vessels. This damage might eventually cause or worsen high blood pressure and other forms of heart disease or heart problems.

Let's take a look at some of these complications and problems.

- High blood pressure—Sudden drops in blood-oxygen levels that occur during sleep apnea episodes increase blood pressure and place a strain on your cardiovascular system. Approximately 50 percent of people with sleep apnea develop high blood pressure (hypertension), which can play a serious role in heart disease.
- Heart failure—Risk of heart failure might increase with sleep apnea because of the swings in blood pressure that occur during episodes. Combined with reductions in oxygen to the heart tissue, it might damage your heart muscle.
- Arrhythmias—Arrhythmias are heart rhythm problems that occur when the electrical impulses in your heart don't function properly. When this happens, it causes your heart to beat too fast, too slowly, or irregularly. In relation to sleep apnea, atrial fibrillation is the most common type of arrhythmia to occur.

- Coronary artery disease—Caused by the gradual buildup of fatty deposits in your coronary arteries, your heart muscle receives less blood as the buildup slowly narrows your coronary arteries. This diminished blood flow may cause chest pain, shortness of breath, and other symptoms. Sleep apnea may increase your risk of coronary artery disease because of the swings in blood pressure that occur during sleep apnea.
- Stroke—When the blood supply to part of your brain is interrupted or severely reduced, your brain tissue is deprived of oxygen and nutrients. Within just a few minutes, brain cells begin to die. Through the damage and stress to your blood vessels, it is believed that this is caused by blood pressure and oxygen changes from sleep apnea, which increases your risk of stroke.

In the last ten years, sleep apnea treatments have improved significantly, which means if you have already tried addressing and treating your sleep apnea before with no luck, try again. Because sleep apnea can increase your risk of high blood pressure, heart disease, and stroke, it is just one more reason for you to seek treatment before it's too late.

Diabetes

If you are not sleeping well at night, you might want to have your blood sugar levels checked by a physician. Patients seeking treatment for of out-of-control blood sugar levels are often asked how they are sleeping. The answer is frequently the same: not so well.

When blood sugar levels rise to abnormally high levels, the kidneys attempt to get rid of it through urination. When this occurs, getting up and going to the bathroom throughout the night can wreak havoc on an otherwise normal pattern of sleep. Diabetes and sleep complications go hand in hand, as clearly blood sugar anomalies can overtly contribute to sleep loss. Evidence even exists which suggests that not sleeping well can increase your risk of developing diabetes.

People who have poor sleep habits are at a greater risk for becoming overweight or obese, which increases their risk of developing type-2 diabetes. Chronic sleep deprivation may also lead to insulin resistance, which can result in high blood sugar and diabetes. It has been found that about 23 percent of men with diabetes were found to have sleep apnea, compared to only 6 percent of the general population.

It has been shown that rates of diabetes are higher among people with sleep apnea, as they share common risk factors, including obesity and advancing age. More than half of people that are obese are considered to be at a high risk for developing sleep apnea. Further studies also suggest that having sleep apnea increases the risk of developing diabetes.

Both diabetes and sleep apnea share a long list of potential complications. As listed previously, these complications include high blood pressure, cardiovascular disease, eye disease, changes in glucose metabolism, and weight-control problems. A person who suffers from sleep apnea can experience difficulty with losing weight due to the constant stress of not sleeping, as it causes high levels of cortisol and steroids, which make weight loss more difficult. Moreover, when you are tired day in and day out, you might not have the energy you need to hit the gym harder, so the cycle continues (albeit the *Spinning* might stop).

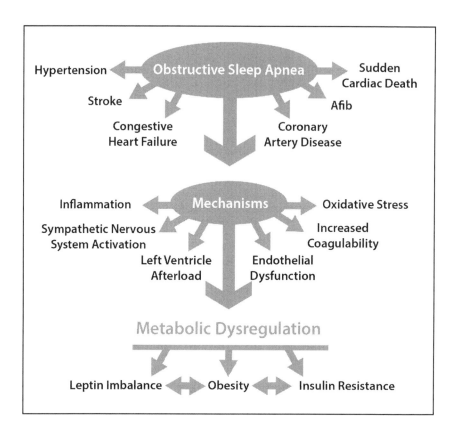

While treating sleep apnea is an important step toward finally getting a good night's rest, treatment can also help address and possibly eradicate many other diabetic complications. In addition to decreasing daytime sleepiness and removing a barrier to effective weight loss and/or management, treating sleep apnea can also improve an array of other complications that we will touch base on further in this chapter, including

- psychological well-being
- memory, concentration, and other cognitive functioning
- erectile dysfunction

- lower blood pressure levels
- productivity during the day, with fewer sick days
- decrease in the risk of traffic accidents

For those with diabetes, seeking treatment for sleep apnea is vital to maintaining good health. Through various research, it has been found that insulin sensitivity improved and A1C levels decreased after only a few months of treatment with CPAP. Speak to your diabetes health-care team, which includes your dentist and ENT, to learn about your options. Getting a better night's sleep through sleep apnea treatment can go a long way toward resolving diabetes complications.

Impotence

While waking up feeling exhausted is tiring in itself, you might have to watch out for something else: impotence, or erectile dysfunction. Some of the more common symptoms of sleep apnea include fatigue, high blood pressure, and weight gain. However, a growing body of research has found that sleep apnea can be a drain on intimacy, causing erectile dysfunction in men and a loss of libido in women. The occurrence of sleep apnea often triggers impotence for younger men, and the findings in a small study suggest that treating your sleep apnea can effectively jump-start your sex life.

Just as sleep is an important part of your life, so is sexual activity. When erectile dysfunction occurs, it can take a toll on your overall well-being. Occasional inability to maintain an erection can be caused by too much alcohol, stress, or a number of other factors. However, a frequent inability indicates a problem that for most men is treatable.

Today, there is evidence that shows that sleep disorders are linked to erectile dysfunction. Throughout the day, but more so during sleep, there are cycles of hormones and other physiological chemicals released into the body. One of these cycles releases testosterone, which is an important hormone for many aspects of sexual health. If you suffer from sleep apnea, you will experience interrupted testosterone cycles. With a disrupted hormonal cycle, it may be the link between your sleep apnea and the ability to maintain an erection.

Treatment with the CPAP machine is often used in the treatment of sleep apnea, but the availability of oral appliances also offer an improved alternative to this machine. By treating sleep apnea with CPAP or an oral appliance, it has been shown that men experience an improved ability to maintain an erection and, by "extension," improved their sex life. Through the treatment of sleep apnea, sexual dysfunction improves dramatically, which helps to enhance energy, mood, and overall quality of life.

Accidents

One of the main symptoms of sleep apnea is excessive daytime sleepiness. Just like excessive speed, alcohol, aggressive driving, and inclement weather, sleepiness often contributes to or causes car accidents. In the past few years, sleep as a factor in car crashes has begun to be investigated. Daytime sleepiness can put people with sleep apnea at an increased risk of falling asleep behind the wheel. People with sleep apnea are also up to five times more likely than normal sleepers to experience traffic accidents.

While many studies indicate that, due to sleep deprivation, young males under the age of twenty-six cause the most fall-asleep car accidents, the number of sleep apnea related crashes still remains unknown. Sleep, like exercise and proper diet, is crucial to good health, but you can't accomplish this when you suffer from sleep apnea, which is why treatment is so necessary.

Sleepiness due to sleep apnea does cause and contribute to accidents. In fact, a higher percentage of crashes caused by a person falling asleep result in fatalities than those attributed to other causes. Due to this information alone, we would want to place a strong emphasis and importance on seeking treatment for sleep apnea. As more attention is paid to these accidents, the steps that can reduce them need to be explored as well. It is our belief that the public outcry necessary to shine the light on the virtual epidemic of sleep apnea will in part be driven by the tragedy of so many lives lost on our roads. If only that awareness could be achieved without the need to reach such a tipping point.

While the increased risk for health problems linked to sleep apnea is scary, receiving safe, effective treatment for OSA is readily available—and necessary. In many cases, treatment might be with a CPAP machine (as we mentioned previously), but remember that there are also other less intrusive treatments options. Oral appliance therapy, as well as quick, virtually pain-free, surgical solutions are new techniques that you and your doctor can explore. Together, discover if they are good options for you to consider in managing your individual condition. Through treatment of sleep apnea, your risk of daytime sleepiness and car accidents is greatly reduced, if not all but eliminated. For this reason, it is important to talk to your doctor about treating your sleep apnea so that you can prevent both accidents and related health problems.

Obesity Hypoventilation

Obesity hypoventilation syndrome is also known as Pickwickian syndrome. The condition occurs when severely overweight people fail to breathe rapidly enough or deeply enough, resulting in low blood oxygen levels and high blood carbon dioxide levels. When this happens, many people with this condition frequently stop breathing altogether for short periods of time during sleep—de facto sleep apnea. This disease puts an extreme strain on the heart, which can eventually lead to heart failure. The most effective treatment for this syndrome is weight loss, but other forms of treatment for sleep apnea can significantly improve symptoms as well.

One way to treat obesity hypoventilation is through the use of CPAP machines, which are used while you are sleeping. Through this treatment, your airways are kept open, which helps you to breathe during the night and eliminate frequent moments of waking up.

It is also recommended that you lose weight as part of your treatment plan. Successful weight loss often involves setting goals and making lifestyle changes, such as following a healthy diet and being physically active. By losing weight, you can eliminate the severe symptoms you have been experiencing with obesity hypoventilation syndrome.

Just like sleep apnea, obesity hypoventilation alone can also lead to other serious health problems. By seeking proper treatment for both conditions, you are helping to protect your overall health and well-being. Your health-care team, treatment modality, and your family can help you manage your treatment while giving you the motivation you need to continue with treatment. The goals of treatment may include

- supporting and aiding your breathing
- achieving major weight loss
- treating underlying and related conditions

Taking the next step toward health involves a successful treatment plan. As previously mentioned in this chapter and other chapters, we cannot be emphatic enough about the importance of treating sleep apnea. Sleep apnea is often the underlying cause for an array of conditions, and the best way to improve those conditions is by addressing their source--through the treatment of your sleep apnea.

So protect yourself by seeking treatment for sleep apnea now, before you develop worsening of symptoms or the development of another serious condition.

Case Study: M and Obesity

M lost weight after surgery, but he didn't have weight-loss surgery; he had sleep-apnea surgery. At five feet, five inches tall, M weighed over three hundred pounds. The surprisingly good news about sleep apnea and weight loss is that weight loss often follows good sleep apnea therapy. According to studies published recently, it appears that sleep apnea may be a big player in causing obesity. Virtually everyone who has sleep apnea has been told that he or she needs to lose weight. However, the relationship may not be as straightforward as simply weight gain adding to the thickness of the neck, resulting in sleep apnea. There used to be a joke about people who are overweight claiming that there was a problem with their hormones. It turns out that in the case of sleep apnea and weight gain, there may be a lot more to that than we would like to think. It's pretty clear from studies focused on weight-loss surgery that the weight loss does not always correlate with the amount of improvement in sleep apnea when present.

Anecdotally, there are also many cases wherein people who lost a lot of weight through diet and exercise did not necessarily improve their sleep apnea. The body has a number of systems that help us to maintain weight along with temperature and just about everything that we do. Just like a house has a thermostat to regulate temperature, the body has systems that help make us stay within certain physical parameters. Your thermostat uses electrical signals linked to the furnace to signal when it needs to turn on and off. Your body also sends electrical signals when something needs to happen very quickly. These signals are carried by nerves.

Aside from nerve-transmitted signaling, the brain also communicates directly with the pituitary gland to generate chemical signals for transmission to the rest of the body. These signals are not intended to act very quickly and are therefore sent by slower messengers. The chemicals float through the blood and attach themselves to their target organs and/or cells. These chemicals are called hormones. Electrical nerve signals would be like telephone call or e-mail and the hormones are like regular mail carrying a court order. The first hormone most people think of when they think of weight gain is the thyroid hormone. However, it turns out that patients with sleep apnea don't experience a higher incidence of hypothyroidism than the general population. This suggests that thyroid deficiency is probably not involved in the development of sleep apnea in the majority of patients diagnosed with OSA. Patients with severe thyroid deficiency can develop a condition wherein all of their skin and the lining of their nose and throat airway becomes thick. This condition is called myxedema. Patients with myxedema are at increased risk for sleep apnea, but this is by no means the majority of patients with OSA. Patients who do have hypothyroidism may worsen the sleep apnea when they first receive replacement thyroid hormone, however. Therefore, the general recommendation is to have them begin some form of sleep apnea therapy before they start their thyroid hormone. But if thyroid hormone is probably not involved in the weight gain associated with sleep apnea, what is?

We have seen that sleep apnea causes low oxygen levels in the blood to the brain and other organs. When this low oxygen is recognized by the brain, the brain then has to wake up enough to address the oxygen problem. The wake-up signal alters the brain chemistry. The chemistry changes alter some of the local messengers in the brain called neurotransmitters. These changes affect the pituitary, also called the master gland. The pituitary generally represents the go-between or middleman between the brain and the other hormone-producing glands. There are exceptions to this rule, such as growth hormone. The growth hormone directly affects almost all of the other cells in the body. It is generally associated with giants and a condition known as acromegaly. The people affected by this problem become extremely tall if the condition presents before they finish growing. Another anomaly that comes along with acromegaly is a large tongue size and growth of the jaw, with both factors contributing to a high incidence of sleep apnea. Growth hormone also causes problems that drive sugar metabolism out of balance. Sleep apnea also appears to cause the release of another hormone called ghrelin, which is produced inside the brain. Ghrelin causes the release of growth hormone. One very interesting fact about ghrelin is that it may also influence adult appetite. The short of it is that ghrelin makes us eat, which tends to produce fat. It also has some influence on insulin resistance.

Atrial Fibrillation (Afib)

Previously, we have discussed sleep apnea in relation to GERD, depression, heart disease, stroke, diabetes, and more. What about its connection with atrial fibrillation (Afib)? Today, a clear connection has been established between sleep apnea and atrial fibrillation. Similar to cardiovascular disorders, there is continuing controversy as to whether sleep apnea is merely a common coexisting condition among patients with AF or whether it is a true causative factor.

Both sleep apnea and atrial fibrillation are very common, with over two million adults in the United States having AF and at least one in fifteen having moderate to severe sleep apnea. There is a greater risk of atrial fibrillation among those with obstructive sleep apnea than even among those with other cardiovascular diseases. Atrial fibrillation patients with untreated sleep apnea are more likely to revert into Afib after electrical cardioversion than Afib patients who do not present symptoms of sleep apnea.

Additionally, when the body does not take in enough oxygen, Afib can lead to heart and valve diseases, sleep apnea, and chronic fatigue. And because many who have atrial fibrillation also have sleep apnea, the importance of treatment is greater than ever. By treating sleep apnea, you are reducing the risk of atrial fibrillation. Given the prevalence of sleep apnea among patients with atrial fibrillation, it is important for individuals to be assessed for both conditions. If sleep apnea is suspected, proper treatment is essential in order to maintain and improve overall health.

Morning Headaches

An array of symptoms can be eliminated through the treatment of sleep apnea, including morning headaches. Headaches can be disruptive to sleep, and when you seek treatment, that which is offered often leads to insomnia or sleepiness, a classic catch-22. The treatment of sleep apnea may in fact help to prevent headaches that awaken one from sleep or that occur first thing in the morning.

If you experience morning headaches, it typically indicates that your sleep apnea is at least moderate, if not severe. When you experience headaches in the morning, it is often due to the significant changes that occur within the bloodstream as a result of airway obstruction. During episodes of sleep apnea, your body experiences changes in oxygen and carbon dioxide levels.

If you block the airway and keep the lungs from moving air in and out, two things will happen. First, the carbon dioxide trapped in the lungs cannot escape, which increases its level within the bloodstream. Second, oxygen is unable to get in and its level in the bloodstream begins to drop. These morning headaches result from the rise in carbon dioxide. As carbon dioxide rises, it causes the blood vessels in and around the head to dilate, which results in migraine-like headaches that are throbbing and irritating.

Morning headaches as a result of sleep apnea usually are experienced on both sides of the head rather than just on one side. These headaches also typically resolve on their own within thirty minutes of awakening. By treating your sleep apnea, you can help eliminate the obstruction in your airway so that you can get a better night's sleep while also waking up without the painful burden of a morning headache.

The relationship between headaches and sleep apnea is complex, but it is important for those suffering chronic headaches to seek treatment immediately. Through the treatment of sleep apnea, you could potentially find relief from your morning headaches and/or chronic headaches.

Sleep and Stress

When we experience stress, it affects us emotionally, physically, and behaviorally. The right amount of stress can be a positive force that helps us to do our best by keeping us alert and energetic. However, too much stress can make us tense, anxious and can cause sleep problems. A list of common signs of stress includes

- depression
- sleep problems
- tension
- anxiety
- work mistakes
- poor concentration
- apathy

We each react to stress differently, but a majority of people will exhibit sleep problems when experiencing stress. By properly managing your stress and your sleep problems, you can get a better night's sleep. As with conservative methods of treatment for sleep apnea, you can try the following to improve your sleep:

- Assess what is stressful.
- Seek social support.
- Practice thought management.
- Exercise.
- Eat a healthy diet.
- Get adequate sleep.

These steps will help many people get a better night's sleep, but if you suffer from sleep apnea, the stress is less likely to fully diminish. Adequate sleep is crucial to proper brain function, just as air, water, and food are important, but stress can alter your sleep-wakefulness cycles. Any amount of sleep deprivation will diminish mental performance, which is why it is important to get the right amount of sleep each night, while also seeking treatment for sleep apnea.

Sleep apnea hinders your mental performance and wakefulness in the mornings. As we have stated previously, seeking treatment for your sleep apnea is essential in maintaining your health, and that begins with your stress level. Watch out for your stress, as it directly reflects your sleep patterns and health.

Testosterone, Cortisol, and Leptin Ghrelin Growth Hormone

Let's begin with a look at testosterone. Recent research has concluded that testosterone is directly affected by quality and quantity of sleep you get. When you suffer from sleep apnea, your levels are significantly affected due to lack of sleep. When we habitually exercise poor sleep habits, it can slash testosterone levels. Furthermore, the relationship between testosterone and sleep is linked to the interactions of testosterone and cortisol.

As stated previously in this chapter, stress plays a significant role in your sleep patterns. To expand on this, cortisol is a hormone that is produced by your adrenal glands in response to stress. Through this, cortisol has been shown to be the problem in sleep regulation for two different reasons:

- the impairment of sleep and promotion of wakefulness
- testosterone and cortisol levels that are inversely proportional

Testosterone levels appear to decline in aging men, while cortisol levels have been found to increase, linked to a diminished quality of sleep. If this occurs, what do you do? Get more quality sleep. We understand that sleep apnea can hinder your ability to get an appropriate amount of sleep, but seeking treatment will greatly improve your sleep patterns, which in turn will improve your testosterone and cortisol levels. It's a win-win situation.

Moving on from testosterone and cortisol, let's take a look at leptin and ghrelin—hormonal control of appetite and body fat. When hunger increases, we tend to eat more, which can lead to an increase in weight. Previously, we mentioned that weight and obesity track with sleep apnea symptoms. The connection is primarily due to leptin and ghrelin, hormones which affect energy balance.

Leptin is secreted primarily in fat cells, as well as the stomach, heart, placenta, and skeletal muscle. It is utilized by the body to decrease hunger, while ghrelin is secreted in the lining of the stomach and increases hunger. Both hormones respond to how well fed you are, with leptin correlating to fat mass. The most important relation between these two hormones is that their signals get messed up with obesity.

As we know, obesity is one of the leading causes of sleep apnea, which means it is important to monitor the levels of both leptin and ghrelin to prevent weight gain. It is also vital to seek treatment for sleep apnea in order to achieve an appropriate amount of restorative sleep each night. This helps to regulate leptin and ghrelin levels, preventing excessive weight gain and subsequent risk of obesity.

PART II

Treatment

Resolve

Decide Firmly on a Course of Action

You don't drown by falling in the water,
you drown by staying there.

—Edwin Lois Cole

The struggle to breathe is fundamentally the struggle to survive. From our first gasps for air outside the womb, the battle begins. Humans and all animals seek first to oxygenate their blood before all else. It is our primary instinct. If you are experiencing symptoms of sleep apnea, that struggle to breathe may not enter your conscious mind, but make no mistake about it: your body is being attacked *nightly* and *persistently* by a silent killer. This sinister killer rarely acts quickly, preferring to carry out the act of murder in such small increments that victims are left unaware that a capital crime is being committed—right in their own homes, while they struggle to sleep.

If you have been diagnosed with sleep apnea, treatment is absolutely essential in order to live life to its fullest and may be necessary, ultimately, to sustain life itself. Nobel Prize-winning philosopher and physician Albert Schweitzer summed up the tragic loss of human potential with these words: "The tragedy of life is not that we die, but is rather, what dies inside a man while he lives."

The time is now. Action must be taken. Now that the crime has been exposed, and the silent killer identified, it is crucial to defend yourself against the nightly assault. This elusive killer will never be apprehended. The only way to protect your most valuable asset is to protect yourself. Just as you would never leave a door or window open if you knew the thief was lurking, this thief of oxygen—this killer—can only be stopped through prevention.

The following chapters outline the various treatments available to keep the killer at bay.

Chapter 8

VOAT:
Ventral-Only Ablation of the Tongue Base

If you are using CPAP to effectively mediate symptoms, great! By working closely with your physician, you have found a solution that works for you and presumably are reaping the benefits of greater energy, greater concentration, and overall more effective living. Congratulations! You may stop reading this book now, but we would

ask that you please help spread the word about sleep apnea to your peers in spite of the competitive edge your successful treatment now affords you.

Unfortunately, you—if you are still reading, that is—are in the minority of sleep apnea patients. Research shows that among CPAP users, there is a disturbing pattern of noncompliance. These patients have the solution to their struggle with the silent killer but leave it at their bedside, perhaps thinking it a deterrent like the security alarm sign in the front yard or, for supporters of the Second Amendment, that personal firearm secured inside your home for self-defense. Frankly, the sign and the gun might provide greater protection from harm than the unused or under-used CPAP machine. To extend the analogy, the gun, like the idle CPAP machine, indeed poses its own threat. It is a dangerous mistake to become complacent or cavalier with your therapy.

It is important to continue to see your physician regularly if you have sleep apnea. The disease is deceptive and may continue to affect your body's chemistry unless the therapy is administered effectively and regularly. If you are having trouble with adapting to CPAP therapy, there are other solutions available to you.

Now there is a revolutionary procedure that has the capacity to eliminate need for CPAP therapy. For many patients, the procedure, ventral-only ablation of the tongue (VOAT) is nearly ideal. It is a minimal invasive, in-office surgical procedure that can be performed in seconds and can provide lasting and almost immediate relief from your symptoms of sleep apnea. Only you and your doctor or sleep specialist can effectively evaluate whether or not it is right for you. Toward that end, let's look at one method for evaluating any medical procedure you might consider. Organizing your thoughts is important for anything in life, but especially when you are going to alter your body, even in small ways. All you have to remember is PEPPER.

The ideal procedure to fix any problem would be

- painless
- easy
- performed once
- permanent
- effective
- reproducible

PEPPER for short. For sleep apnea, this doesn't exist—**yet**. But we have gotten the painless, easy, effective, and reproducible parts down.

How do we improve the opening in the throat while not compromising its ability to keep doing what it needs to do? The tongue is probably the most important structure in the mouth. It is also the structure that most often cuts off the airway at night. Simple enough: let's cut it down to size. One problem though is that we need our tongue and throat to swallow in order to survive. So we have to be careful when we operate on it because swallowing (and talking) are important.

Pain and Painless

Let's face it: most of us avoid pain and associate pain with surgery. As a result, we tend to avoid surgical solutions, even when that can clearly benefit us. We have talked about the tongue causing the lion's share of decreased airflow during sleep apnea. That is because the tongue represents the largest movable soft mass in the throat. Now, cutting the tongue is painful mostly because of nerve endings in the lining (known as the mucosa, a kind of a wet "skin," if you will). The muscle does have nerve endings, but not as many.

In 1999, Dr. Nelson B. Powell published a paper about using a new technology to improve snoring and sleep apnea. This was followed by a series of papers on the technique. While the technique would have sounded like science fiction fifty years ago, it is amazingly simple in concept: use a radio wave, a form of electromagnetic energy like light, to heat up the tissue and shrink it. This is kind of like using a microwave to warm up things. It turns out the procedure he described was less painful than the standard procedure for sleep apnea. *A lot less painful.*

Now some form of electromagnetic wave has been used in medicine since before Babe Ruth got radiation for cancer. We all understand that X-rays are a high-energy form of light we use for diagnosis, and radiation helps us with cancer treatment. What you may not be familiar with are electrical devices to coagulate blood. Now these aren't new. Dr. Cushing, the first neurosurgeon, helped discover the use of electricity to cauterize blood vessels early in the twentieth century.

It has long been known that we could use this to reduce the size of the lining in the nose. For more than seventy-five years, we have routinely been using an electromagnetic wave to change the proteins in blood, changing the form of blood from a liquid to a solid. The resistance to the radio wave passing through the body's tissues heats the tissues up. (Again, think, *Microwave oven.*) This changes the configuration and, in the particular case of the tongue, causes it to actually shrink.

Having established earlier in the book that "wider is way better," the amount of room we have to breathe has a great deal of influence on how much air we are able to get into our lungs. The nose and the sinuses are part of the pipeline supplying air to the lungs. One of their important functions is to warm and moisten the air we breathe. This makes the air very thin and easy to move into the lungs. But the nose from breath to breath does not change as much as the tongue and soft palate (roof of the mouth). The soft palate and tongue can move in and out of the air stream to close off the airflow. The palate has received a lot of attention in the past. People have had artificial stiffeners installed or had parts of the soft palate removed (UPPP or uvulopalatopharyngoplasty). This has met with some success in reducing snoring, but less success in reducing sleep apnea. Pointedly, it is a real disappointment as an effective therapy for sleep apnea. It is painful. Some people have told me it was worse than childbirth. Others have told me it was worse than open-heart surgery.

It also comes with the disagreeable problem of complicating the process of swallowing. To be specific, it tends to cause food to get into the nose when patients swallow. The palate is designed to keep food out of the sinuses. Most patients don't enjoy using their nose as a peashooter …

Now the reduction in sleep apnea for the UPPP procedure is limited. Why? Our theory is that it doesn't have as much bulk as the tongue and therefore doesn't have the efficiency of the tongue at closing off the windpipe.

The Loose Hammock and the Swallow

The tongue is like a hammock in a lot of ways. It is soft and is strung between two points to provide support. The tongue complex is much larger than you think. It is basically almost everything that is muscle in front of the throat and between your lower teeth and jaw. (For those of you inclined to look this up, Google *genioglossus, geniohyoid, mylohyoid, stylohyoid,* and/or *digastric muscle,* and that should give you an idea.) It is supported by the jaw and the bone at the base of the skull below and behind the earlobe or mastoid.

If you have seen the kind of hammock that goes on a portable stand and lifted one end up, you will have seen that gravity shifts the bulk of the hammock to the lower support pole. Now if you loosen the hammock, even more of the hammock hangs toward the lower pole.

This is what happens to the tongue when we relax it. When we open our mouth and the front of the jaw moves down and backward (try it yourself), the jaw moves closer to the skull base and the tongue (hammock) gets looser. When we lie down, it's easy to see why it can close off the airway.

Everyone knows you can't swallow and breathe at the same time. If you didn't know this, just try it! The tongue is designed to help close the epiglottis. The epiglottis is part of the voice box and the structure that helps close off the windpipe when it is needed. The front part of the tongue pulls on a ligament called the hyo-epiglottic ligament, which pulls the epiglottis and voice box forward. The rest of the tongue complex then pushes down on the epiglottis. This creates an effect like a cheap magician trick where the plastic flowers are pulled into a hollow wand on a string.

The vocal cords also closed at this point and the whole thing is sealed off very nicely. It is a neat trick and protects precious territory, because your lungs have to last your entire life. Generally speaking, you can't just get a new pair.

Now there is a fair amount of pressure generated when you swallow, so you can imagine that this system is pretty effective. Sometimes, it is too effective. The tongue gets so much exercise that it gets big. When the tongue gets big (or loose), it can be more likely to fall into the airway when we are lying down. (Apparently, talking does far less to exercise the tongue than does eating, or the incidence of sleep apnea on women might be considerably higher!)

One of the things you always see discussed is that people need to lose weight. While this is helpful, it is not shown to have a direct causal effect on the sleep apnea process. The assumption some people have is that the tongue actually gets fat. This is absolutely untrue. The tongue is a muscle, through and through. It never gets fat. Surgical procedures on the tongue reveal no presence of fat. However, the tongue does seem to get larger for some reason, and many theories are out there attempting to explain the growth, but none have been validated.

People have also heard of people swallowing their tongue. It is a myth that people having a seizure are swallowing their tongue. However, the tongue does passively close the airway and cause problems in these patients as it hangs backward like a loose hammock. It is almost exactly the same mechanism that happens in sleep apnea.

The Standard

So if we are going to decide on a treatment plan for a sleep apnea patient, what it the ideal procedure sequence? Again, it should be painless, easy, performed once, permanent effective, and reproducible: PEPPER.

One of the advantages of the new technique described by Dr. Powell is to get into the muscle with a needle and to be able to alter structures by shrinkage. Most people have seen bacon cook and know that it shrinks. All imagery of tongue on a skillet aside, this is one way of looking at what happens in this technique.

Now let's see how the radio frequency ablation method for tongue volume reduction stacks up to the PEPPER test.

Painless

It is performed through a series of puncture wounds at the base of the tongue. It does stack up to the painless test much better than any of its predecessors. It's predominately performed within the muscle of the tongue and not in the lining mucosa, therefore away from most of the nerves. The theory does seem to hold up, because the patients return to normal diet and stop taking pain medicine more than ten times faster than the UPPP patients. Now in terms of risks, the bleeding risk is all but eliminated. Although swelling can be a bit of a risk, especially in the apnea patient, who is prone to collapsing the airway, it is minimal. The procedure is painless enough that it has been adapted by some doctors into an office procedure under local anesthesia.

So in summary, while not perfect in terms of pain, most people who choose VOAT experience less pain during and after the procedure than patients undergoing nearly every other current procedure.

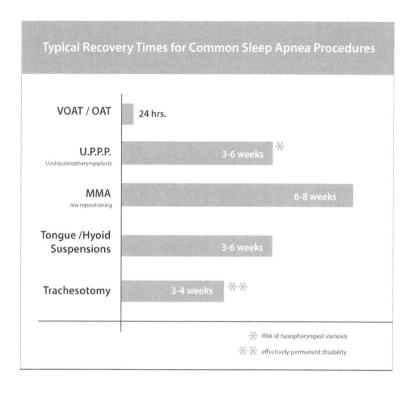

Typical Recovery Times for Common Sleep Apnea Procedures

VOAT / OAT	24 hrs.
U.P.P.P. Uvulopalatopharyngoplasty	3-6 weeks *
MMA Jaw repositioning	6-8 weeks
Tongue /Hyoid Suspensions	3-6 weeks
Trachesotomy	3-4 weeks **

* Risk of nasopharyngeal stenosis
** effectively permanent disability

Easy

The procedure is typically performed by otolaryngologists. These doctors are trained to do cancer surgery of the head and neck, along with complicated ear and nasal surgery. This procedure is technically one of the easiest procedures they will ever do.

Performed Once

Here the procedure falls short. It is designed as a series of up to five or more treatments to reduce the tongue. This is in no small part an effort to ensure safety because the tongue is so vital to swallowing and speech. If we reduced it too much, one or both of these functions of the tongue might be impaired.

Permanent

Well, sort of. All forms of apnea surgery seem to have a tendency to not be permanent. Even surgeries that move both upper and lower jaws forward, although very effective initially, tend to fail at the five-year mark after surgery. The radio frequency ablation of the tongue procedure, however, precisely because it is essentially painless and designed to be repeated, does have an advantage here. You can repeat it with no ill effect.

Effective

The following is the position statement of the American Academy of Otolaryngology, Head, and Neck Surgery (the oldest medical and one of the most respected specialty associations in the country) regarding radio frequency ablation of the tongue.

> Adult patients with mild to severe obstructive sleep apnea (OSA) can be successfully treated with submucosal radiofrequency tongue base ablation. (Powell 1999 and refs below) The majority of studies demonstrating effectiveness of tongue base submucosal radiofrequency ablation (RFA) have been performed in patients with mild to moderate OSA and without morbid obesity, often as part of multilevel pharyngeal surgical therapy.
>
> A randomized, CPAP and placebo controlled trial of tongue base and palate (RFA) for mild to moderate OSA demonstrated significant improvements following radiofrequency compared with sham-placebo in quality of life (QOL), airway volume, apnea index, and respiratory

arousal index (all P < 0.05).(Woodson 2003, Level 1 evidence). Comparison between CPAP and RFA showed no significant differences in improvements in QOL or daytime sleepiness. (Woodson 2003).

Additional prospective study comparing CPAP to radiofrequency submucosal ablation for mild to moderate obstructive sleep apnea showed similar effectiveness of both therapies suggesting a role for primary treatment of mild to moderate OSA with submucosal ablation (Ceylan 2009, Level 2 evidence).

Prospective study with extended follow-up of patients treated with RFA demonstrates persistent improvements in daytime sleepiness and OSAS-related quality of life (both P < 0.001). Median reaction time testing and apnea-hypopnea index (AHI) were also significantly improved at long-term follow-up (P = 0.03 and 0.01). (Steward 2005, Level 2 evidence).

Cumulative meta-analysis of submucosal RFA found a 31% reduction in short-term ESS (odds ratios (OR) 0.69, 95% confidence interval (CI) 0.63-0.75), which was maintained beyond 12 months (OR 0.68, 95% CI 0.43-0.73). Likewise, RFA resulted in a 31% reduction in short term (<12 month) (OR 0.69, 95% CI 0.61-0.77) and 45% reduction in long-term (>24 month) (OR 0.55, 95% C.I. 0.45-0.72) respiratory disturbance index (RDI) levels. (Farrar 2008, Level 3 evidence).

Randomized comparison of submucosal RFA with tongue suspension found significant improvements in OSA for both treatment groups with significantly less morbidity with submucosal ablation. (Fernandez-Julian 2009, Level 1 evidence). Other studies have also demonstrated low morbidity with tongue base \ submucosal RFA. (Kezirian 2005, Level 4 evidence).

Controlled study of treatment schemes for RFA suggests additional improvement in outcomes with repeated treatments. (Steward 2004, Level 2 evidence). However, more recent studies have demonstrated significant improvement with a

single high energy treatment session with low morbidity. (Nelson 2007, Level 4 evidence).
Adopted 5/3/2010
Reaffirmed 12/8/2012

Important Disclaimer Notice

Position statements are approved by the American Academy of Otolaryngology—Head and Neck Surgery, Inc. or Foundation (AAO-HNS/F) Boards of Directors and are typically generated from AAO-HNS/F committees. Once approved by the Academy or Foundation Board of Directors, they become official position statements and are added to the existing position statement library. In no sense do they represent a standard of care. The applicability of position statements, as guidance for a procedure, must be determined by the responsible physician in light of all the circumstances presented by the individual patient. Adherence to these clinical position statements will not ensure successful treatment in every situation. As with all AAO-HNS/F guidance, this position statement should not be deemed inclusive of all proper treatment decisions or methods of care, nor exclusive of other treatment decisions or methods of care reasonably directed to obtaining the same results.

Now let's examine further what the academy has had to say about how the effectiveness of RFA. For the nerds in the crowd, here are the bullet points and references extracted from the American Academy of Otolaryngology, Head, and Neck Surgery position statement on the procedure:

- Comparison CPAP/ RFA no significant differences QOL or EDS. (Woodson 2003).
- Prospective study CPAP/RFA mild/mod OSA similar effectiveness (Ceylan 2009, Level 2 evidence).
- Prospective study with extended follow-up RFA persistent improvements EDS and OSAS-QOL (both P < 0.001).

Median reaction time testing / apnea-hypopnea index (AHI) significantly improved at long-term follow-up (P = 0.03 and 0.01). (Steward 2005, Level 2 evidence)

- Meta-analysis RFA 31% reduction ESS (odds ratios (OR) 0.69, 95% confidence interval (CI) 0.63-0.75), maintained beyond 12 months (OR 0.68, 95% CI 0.43-0.73, (Farrar 2008, Level 3 evidence).

- RFA resulted in a 31% reduction in short term (<12 month) (OR 0.69, 95% CI 0.61-0.77) and 45% reduction in long-term (>24 month) (OR 0.55, 95% C.I. 0.45-0.72) respiratory disturbance index (RDI) levels. (Farrar 2008, Level 3 evidence).

- Randomized RFA with tongue suspension found significant improvements in OSA for both treatment groups with significantly less morbidity with submucosal ablation. (Fernandez-Julian 2009, Level 1 evidence). Other studies have also demonstrated low morbidity with tongue base/ submucosal RFA. (Kezirian 2005, Level 4 evidence).

- Controlled study of treatment schemes for RFA suggests additional improvement in outcomes with repeated treatments. (Steward 2004, Level 2 evidence) However, more recent studies have demonstrated significant improvement with a single high energy treatment session with low morbidity. (Nelson 2007, Level 4 evidence).

- Mild to severe (OSA) successfully treated RFA. (Powell 1999).

- Majority studies mild to moderate OSA RFA without morbid obesity, part of multilevel pharyngeal surgery.

- A randomized, CPAP and placebo controlled trial RFA mild-moderate OSA significant improvements (QOL), airway volume, apnea index, and respiratory arousal index (all P < 0.05).(Woodson 2003, Level 1 evidence).

If you would like to review all the literature and judge for yourself, we have included the references in the references section at the end of this book.

The general consensus of the papers is that the treatment of the tongue base as it was originally described relieves about 30 percent of the apnea score in mild to moderate sleep apnea with approximately five separate treatments over several months.

Ventral-Only Ablation of the Tongue Modification

We had been using the standard radiofrequency ablation of the tongue technique for several years when we made a discovery. As the saying goes, everyone can teach you something. We had a patient who had severe obstructive sleep apnea. She had all kinds of medical problems that were related to sleep apnea but just could not tolerate the CPAP mask. She was looking at a tracheostomy procedure (that is the breathing hole in the neck) as an option if she couldn't get her sleep apnea under control.

The patient was desperate. We were going to have to try something "outside the box." We got to the operating room and found that her tongue was so large that we couldn't do the radio frequency tongue reduction procedure the way I had been taught. So I remembered back to my earlier training, recalling a particular cancer operation, and inserted the ablation wand *underneath* the tongue. Surprisingly, the results were better than what I had expected. The patient didn't have as much pain as was typical with the standard procedure. She also didn't have as much swelling and started feeling better sooner than any of our other patients. When we tested her improvement, the results were better than the other patients' had been. As a result, we started trying the technique with more of our patients.

It turned out that the results didn't seem to follow a trend, and our informal observations were confirmed when we started tracking our patients' progress. We were amazed to discover that the results were nearly twice as good as they had been previously. There was less bleeding, and the patients seemed to be recovering faster. We eventually stopped putting people to sleep altogether. We didn't need to. They just really didn't have a much in the way of swelling or bleeding, and whatever bleeding they did experience was very close to the front of the mouth and therefore easier to address and control. Over the next five years, we began to play with the amount of energy we delivered as well as the number of locations that we use to deliver the energy. As often as we could, we would encourage patients to have this ventral-only ablation of the tongue performed as a standalone procedure. The success was so consistently superior that, after a while, we started figuring out that we no longer had to test patients after every procedure.

I started thinking about why the new "ventral-only" approach was working better than the previous procedure. We figured out that this procedure was actually pulling the tongue forward, away from the back wall of the throat. That way, the throat was actually getting wider just behind the tongue. We think that at night this is actually keeping the throat from collapsing. The results seem to be a good deal better. Only one out of every twelve or so patients does not respond to the treatment. Considering the treatment group as a whole, the overall improvement approaches 70 percent as a group with fewer treatments than what had previously been reported. Again, we think that it is because the tongue is pulled in a different direction than that which results from the previous procedure.

The original procedure had only about half the pull in the front to back direction. The ventral approach (VOAT) seems to have almost all its pull from front to back

Oh, and by the way, the original patient never needed a tracheostomy.

So the procedure was working fairly well and the patients were getting better, and we started studying things a little bit more. Turns out that most of the time, change in the tongue doesn't occur immediately. In most cases, the greatest improvement in sleep apnea frequency and severity doesn't improve immediately either. Most patients start to see an improvement in the first couple of weeks, with the most marked improvement being realized roughly between weeks 2 and 8 post-procedure. The data we were collecting started to reveal instances where the apnea recurred in some patients after a period of roughly four years or so. We treated several patients who had been completely cured but after several years began snoring and feeling tired again. We discovered that we were able to treat them again and that their sleep apnea went away, again. We also discovered that, like the original procedure, there was very minimal swelling, very little pain, and very little risk of any nerve damage.

Let's apply some numbers to this in a rough fashion. I've done probably more than five hundred VOAT procedures over the last six years incorporating this technique, with a number of the procedures obviously being repeat treatments due to the nature of the therapy. On average, one out of every twenty will have pain medication requirements lasting longer than twenty-four hours. Only one patient has ever experienced long-term tingling at the tip of the tongue. No patients have had any problems swallowing. One patient did experience a temporary problem with speech, but the effect was very minor and only lasted for about a week. We've elected to admit five patients as a precaution to monitor them for heavier than normal bleeding, and among these, none had any significant blood loss, experiencing only a small amount of discoloration underneath the tongue. That's not to say that you should take any procedure like this lightly, but in general seems, it is no worse than any other oral cavity procedure you might have performed by a dentist or oral surgeon.

Mild sleep apnea is generally defined as an apnea hypopnea index (AHI) between 5 and 14.9. Moderate sleep apnea is between

15 and 29.9 AHI, and severe is anything greater than 30. However, there are many debates on how sleep apnea should be staged. The use of apnea hypopnea index as a definition itself has been debated as opposed to the respiratory disturbance index (RDI). A lot of terms are bounced around, and we will try to give you a no-nonsense approach to understanding the major ones. It is important to be an advocate for your own health care, and it helps to understand whatever you can about your disease(s) and treatment options. As doctors, we treat a lot of people, and I appreciate having patients who are well informed. An informed patient is generally a responsible patient and usually has a better compliance rate and better outcomes. The bottom line is if you don't understand what your doctor is saying, ask him or her. Look up the problems online. It is your health. Own it!

Reproducible

The final PEPPER test is whether or not the procedure is reproducible. Put another way, can it achieve reliable results with multiple patients? The sheer number of references, and the number of individual authors above, should suggest to the person who has not done this procedure that it is reproducible. It can be controlled by power setting, by depth, and by angle of approach.

As a general rule, surgeries develop very slowly and over a long period of time. This type of procedure, being reported only about fifteen years ago, is in its infancy. It is difficult in general to standardize a procedure. For example, tonsillectomies have been around for nearly two thousand years, first recorded to have been performed by Celsus in 30 AD, and people are still writing articles about it.

Let's take a look at the origin and development of the very young science of radio frequency ablation of the base of the tongue (RFA

BOT) for sleep apnea. The radio frequency ablation technique is also variously known as radio frequency volumetric reduction of the tongue base, and my favorite version of this procedure is the ventral-only ablation of the tongue (VOAT). The early versions of the procedure were and still are typically carried either under general or local anesthesia (meaning you are completely asleep, mildly sedated, or fully awake with numbing injections) at the discretion of the surgeon. The wands used to perform the ablation of tissue produce radio waves that circulate around the tip of the needle (which may be in pairs). The waves passing through the tissues heat the muscle of the tongue to roughly forty to seventy degrees Celsius or 100 to 150 degrees Fahrenheit. That is less than boiling but still pretty hot. The needle is placed at the base of the tongue at about the point where it meets the back of the palate and angled downward into the tissue of the tongue. About four to five lesions are usually produced, with the amount of energy delivered being controlled either by time or by measurements within the machine that can define exactly how much energy (heat), measured in joules, is delivered.

Now some background on how pre-surgical testing is used quantify the relative severity of a patient's sleep apnea. Data collected by an array of sensors measure breathing efficiency during a sleep study. Indexes commonly referred to AHI and RDI are derived from the data gathered using a formula to provide a ratio that that yields a definitive diagnosis of sleep apnea, its type, and an accurate measurement severity. During a standard sleep study, many things are measured. Brain wave measurements tell us when you are asleep. Once you are asleep, a little pressure monitor under your nose along with a small thermometer tell us when and how much air you are breathing. An EKG monitors your heart. An oxygen sensor tells us when your blood has enough oxygen in it and when you really need to breathe. A pair of little elastic belts with electrical sensors tell us when your chest and stomach are moving. Sometimes, there are electrodes on your limbs to indicate when you are moving. If all else fails, during a classic study, there is a trained technician to watch, take notes, and make sure you are safe.

In simple terms, when your airflow is decreased by a certain amount and the blood oxygen falls below a certain percentage, we assume you should be taking a breath. Depending on how much your breathing decreases, this event is called an apnea (no breath), hypopnea (small breath), or RERA (respiratory disturbance). The raw data is applied based one of several algorithms developed by different groups of doctors over the years and divided by the number of hours you slept. That is how we get the Apnea Hypopnea Index (AHI), or alternatively the Respiratory Disturbance Index (RDI), which sounds a lot fancier than it is. In short, if your sleep study reveals that apnea occurs while you are not making any effort with your chest or belly, we assume that you are not trying to breathe. When the brain is not signaling your respiratory system to breathe, the diagnosis is central sleep apnea. If you are trying to breathe and cannot, the diagnosis is obstructive sleep apnea. Easy enough in concept.

VOAT: Less Invasive with Fewer Side Effects

One of the most troubling things in a surgeon's life occurs when a procedure he or she has performed on a patient has not been as successful as anticipated, or worse still has failed. This actually happens more often than we would like to admit. It makes life substantially less troubling, both from the patient standpoint and doctor's standpoint, when the procedure doesn't produce a lot of bad side effects. It's always better if the patient does not have to go through a lot of pain and suffering, particularly in instances when such suffering ultimately proves unnecessary, as is the case when a procedure fails. The decision to use the VOAT procedure is a lot easier than some other procedures we have used in the past because of this.

Currently, the only major drawback to the procedure is that there is a bit of a financial risk because some of the insurers have not yet begun to pay for this—a vexing issue of dreaded insurance coding. Fortunately, Medicare does pay for this, which is beneficial because a substantial number of our senior citizens have sleep apnea. Virtually six patients out of every ten over the age of sixty have significant sleep apnea. The other fortunate thing that is good about Medicare covering this tongue reduction is that the procedure tends to help reduce the need for blood pressure medications, diabetes medicines, and cholesterol medications. Since a large number of the elderly are living on fixed incomes, this value-added outcome lowers the overall burden of health-care cost for these patients. Treatment for sleep apnea also helps reduce risk for heart rhythm problems, heart attacks, and strokes. An unusual, surprising side effect for some patients is that it also reduces the need for erectile dysfunction medication.

Unfortunately, there is a failure rate for all surgery. The good news is that the VOAT and radio frequency ablation of the tongue procedures both have relatively low failure rates. For the VOAT, there is a failure rate of roughly one in ten patients. Every surgeon has to develop an algorithm, or decision tree, for how to approach these patients. The order in which sleep surgeries are approached is not widely standardized. Here is our model for evaluating treatment options:

The first step in any successful surgery is correct diagnosis. We have a saying in our office, when things are not going as well as we would like them to go, that "the condition is either underdiagnosed or undertreated."

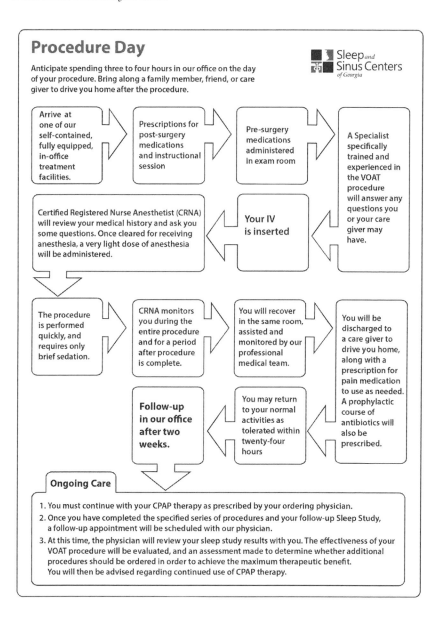

What to expect on the day of your procedure.

Underdiagnosed or Undertreated

If someone's not getting better after visiting the doctor, there are only a couple of possibilities: there is either the wrong diagnosis—or more diagnoses that need to be addressed adding complexity to the problem—or the treatment they are receiving is not adequate for the diagnosis or diagnoses. When looking at sleep apnea, there are a number of things which tend to be underdiagnosed.

In our practice, we find that nasal problems, such as septal deviation and inferior turbinate hypertrophy, sinusitis, nasal valve collapse, adenoid hypertrophy, allergies, and reflux of stomach acid into the nose and voice box, are the most commonly underdiagnosed problems. Of course, unless you happen to be the physician, you probably have no idea what most of the problems listed are—save for allergies, which I am afraid most of us are quite painfully aware of. So let's start at the top and go down and explain them, giving you an idea of what we are dealing with. Let's start with septum deviation.

First, a little etymology. *Septum* comes from a Latin word for *fence*. Clearly, even the Romans understood the role of the nose as protector of the lungs. (A Hellenic scholar might argue that they stole that too from the Greeks.) The nose is actually an enormous structure when you consider the sinuses and all their parts. The nose is divided into two sides: right and left. The two sides are separated by a wall that runs their entire length. That doesn't sound like it runs very deep into the head, but the nose is actually a lot longer than most people think.

Everyone knows where the tip of the nose is, but the back of the nose is actually located roughly between, or slightly in front of, the ears! Considering that the top of the sinuses extend beyond the top of your eyes and the floor of the sinuses are somewhere and just above your molar teeth, the nose and its sinuses are very large indeed. The nose is a very large structure because it performs some very important functions. It warms up the air to keep the lungs from having to waste energy doing it, it filters the air to keep the lungs from being damaged, it helps fight off infections, and it helps to train white blood cells to do their job in combating infection. It helps keep us safe from noxious and poisonous chemicals and keeps us from eating rotten food. It also helps to regulate airflow into and out of the lungs and produces hormones that help control the airflow within the lungs. It also helps conserve water vapor and recover water from the breath. All in all, it's a fairly complicated apparatus. However, its role in causing obstructive sleep apnea is pretty simple.

Back to our garden hose. When the septum is excessively crooked, airflow doesn't pass over it very well. While it's normal to have a little bit of septal deviation, because no septum is absolutely straight, too much misalignment needs to be addressed to allow optimal breathing. In the same way that you streamline a car to make it aerodynamically favorable, the septum needs to be reasonably straight. If you can't get air through the front part of the nose, it's like a kinked garden hose—not much fluid (air) can flow through it.

The turbinates are also major structures within the nose that direct airflow. When you look inside the nose, they look like two small balls of flesh off the side wall of the nose. Really, they are shaped more like cigars. They run from the front of the nose to the back of the nose, parallel to the roof of the mouth. Another reasonable analogy is that they look like the extra fuel tanks or missiles hanging under the wings of a fighter jet. Their job is to direct airflow, and they can sometimes become enlarged, restricting that airflow. They seem to be enlarged in people who have chronic nasal Inflammation for a variety of reasons. When they are large, airflow through the nose becomes much more difficult, so the septum and turbinates can be areas of potential improvement in treating obstructive sleep apnea.

The procedures to correct these two items are septoplasty and inferior turbinate reduction. The pair of these procedures can be done together or individually, and the procedures themselves, in general, have improved dramatically over the past fifteen to twenty years. The septoplasty generally involves an incision inside the nose and either remodeling or removing the deviated portions of the septum. The pain after this type of procedure is generally mild to moderate, and most people don't require pain medication after the first week. Stuffing to prevent bleeding and to hold the septum in place after surgery generally lasts about two weeks. You can usually tell if this surgery is going to be successful if you are breathing about as well as you were before the surgery after the second week. Some of the newer techniques and some of the newer materials we use don't usually require us to pack the nose after these types of procedures. However, if your nose was packed after a septoplasty, that's not to say that the surgeon made a mistake. Some people are just more prone to bleeding. In our practice, we find that the newer procedures eliminate packing in 80 percent of patients. Another practice that we've gotten away from is splinting the nose. Splints are plastic sheets that are sewn inside the nose and may or may not have tubes sticking out of them. The newer techniques have made splinting unnecessary in most cases.

Turbine reductions have likewise gotten a lot better and a number of new tools have been produced for this in the last ten to fifteen years as well. The point here is that the whole process of sinus surgery has gotten a lot easier and really isn't anything you need to be afraid of.

Another condition that is often underdiagnosed is nasal valve collapse. Doctors lacking sufficient otolaryngologic training might not even know what the nasal valve is! The nasal valve is the area of the soft part of the nose that's on the outside the face where it meets the bone. By forcefully sucking in air through the nose really hard, one can temporarily collapse the nasal valve. Patients whose noses don't have enough cartilage, or whose cartilage has become loose or relaxed at its attachment to the bone, experience the collapse of the nasal valve under normal conditions.

Nasal valve collapse can be surgically corrected. In lieu of surgery, the utilization of spring-like products that help open up the nasal valve offer a very simple fix. Breathe Right strips are an example of such products, with many similar products available over the counter or online. If you prefer a surgical solution, be aware that there have been a number of surgical techniques described for improving nasal valve collapse. As always, it is important to find a surgeon who has performed a large number of nasal valve repairs successfully. We generally do not recommend artificial nasal valve implants, which we observe to be suspect in their effectiveness and, moreover, have been known to require removal.

The problem in some people's noses is that the structure is too narrow at the bony opening of the face. Surgical correction for this is the only solution that we are aware of. The procedure to correct this structural anomaly is challenging and needs be performed by an experienced surgeon.

Still another commonly underdiagnosed condition is sinusitis. Sinusitis and sleep apnea tend to occur together with relatively high frequency. Many patients are not even aware of the reduction in airflow attributable to their sinusitis. Our experience suggests that almost one in three patients who describe no sinus problems show evidence of sinusitis, or sinus information, on CT scans of the sinuses. The presence of polyps, or polypoid inflammation of the sinuses, is another fairly common finding that is frequently discovered among patients with sleep apnea. The bottom line in this: if you have an intolerance to your CPAP, or you have newly diagnosed sleep apnea, there needs to be an evaluation for sinusitis and polyps. The evaluation should include a CT scan and a nasal endoscopy. No one should be frightened of Nasal endoscopy. Simply a small "camera" and light, the endoscopy allows us to look closely into the nose and to evaluate the openings to the sinuses. It's generally not painful at all. If you are diagnosed with chronic sinusitis, you should probably be evaluated for allergies and possibly reflux or immunodeficiencies in an effort to determine what triggers the inflammation.

Sinus surgery has come a long way in the last ten years. One of the biggest advancements in sinus surgery finds its origin in the science of cardiology. Just like balloons have allowed cardiologist to resolve clogged coronary arteries, balloons are now being used to help open up the sinuses. The results have been impressive, and recovery time almost nonexistent for most patients. The procedure is so simple and noninvasive that it can be done in the office even under local anesthesia. We have performed hundreds, if not thousands, of the procedures since the technology was introduced. In our experience, it has been a very easy recovery for most patients. If the duration of post-procedural pain or discomfort is used as a gauge, recovery time can be as little as twenty-four hours. Another tremendous advantage of this procedure is that is does not require packing. Packing refers to any form of sterile gauze or sponge material that is placed inside the nose to reduce bleeding. Our experience has been that we have never had to use packing after performing one of these procedures. We offer sedation for this type of procedure, but it can be performed under local anesthesia.

Most people, themselves not being doctors, have only a limited understanding of the structure and role of the sinuses. Sinuses are a series of hollow spaces in the skull on either side of the main passage of the nose. They are kind of like a series of steam rooms along the main corridors of the nose which leak steam into the air as it passes through, humidifying the air we breathe en route to the lungs.

One other structure that can be a problem in patients who have sleep apnea can be tonsils and adenoids hypertrophy. Now the tonsils are not usually missed, but the adenoids are like tonsils in the back of the nose above the soft part of the roof of your mouth. Adenoids can only be seen via nasal endoscopy, or through X-rays of the nose. Inflamed adenoids can causes significant obstruction of the nose, and they are a frequent cause of recurring sinus infections. The adenoids and tonsils are actually similar tissues. They represent a structure that helps mature white blood cells into infection fighting machines. Unlike a tonsillectomy, removal of the adenoids is not particularly painful.

The two other frequently overlooked processes that contribute to overall inflammation inside the nose are allergic rhinitis and chronic rhinitis due to reflux. Reflux is simply acid coming up from the stomach into the tissues above the airway and is increasingly being identified as a factor in upper airway problems. Reflux can also affect the voice box, causing the tissues over the nose and voice box to become swollen and present almost asthma-like symptoms. Think of reflux as repetitive little episodes of vomiting into the nose and throat. When swelling occurs inside the nose, it occurs inside a bony box and there's no place for the swelling to go but inward. The result is a nose that feels like a clogged drain, with nothing being able to move in or out of it.

When recurring reflux impacts the voice box, a number of symptoms occur. The voice tends to get hoarse. The tissues that lie over the windpipe tend to get floppy and can make a noise when sucked into the windpipe. Understandably, this symptom is often mistaken for asthma. Reflux present in the voice box can cause spasms as well. Chronic inflammation can cause a repositioning of the covering to the voice box, causing it to become predisposed to obstruction, leading to an increased risk of apnea. The effects of this repositioning can unfortunately be permanent.

Allergies can contribute to swelling throughout the airway, including the lungs and in the nose. The existence of allergies probably represents an adaptation in immune response by humans to keep out parasites. Most commonly, we think of seasonal allergies such as hay fever. However, asthma and chronic inflammation inside the sinuses can arise from and be aggravated by exposure to year-round environmental allergens.

This list has been provided as a resource to underscore our emphasis that if your sleep apnea this is not improving with treatment, there are things we can look at proactively to improve your situation and outcome with minor surgeries or medical treatments to augment or replace ineffective therapies. If you have been struggling with your current treatment regimen, we hope this chapter serves as a helpful guide for a meaningful discussion of your therapeutic progress with your doctor.

Chapter 9

Dental Solution: Oral Appliance Therapy

Think of oral appliance therapy for snoring or sleep apnea as if you were performing CPR. One of the first things you do is open up the airway by bringing the lower jaw forward. If you can picture this, then you can better understand exactly what an oral appliance does while you sleep. Through oral appliance therapy your airway is opened to help prevent blockage (obstructions) caused by your tongue falling in the back of your mouth/throat.

Oral appliances are custom-made to fit your mouth in order to provide proper management of sleep apnea in each individual. A custom-made oral appliance properly fits your mouth in order to eliminate the unnecessary risk of excessive material. If your oral appliance contained excessive material, it would take away room for your tongue, which would make your mouth even smaller. Typically, there is an upper and lower component that grabs onto your teeth—similar to a mouth guard in sports. Through this, your lower jaw is brought forward to enable us to open up the airway and improve your breathing while sleeping.

Picture yourself wearing a mouth guard or bleaching trays, or even a night guard. These various guards and trays are created in the same way that oral appliances are, while also maintaining the same common goal: to improve function and protect your mouth. Each appliance previously mentioned provides you comfort, which is exactly what this sleep apnea oral appliance will feel like, too. The only significant difference between these devices is that there will be two pieces that attach to each other. As stated previously, this attachment allows the lower jaw to be brought forward and greatly assists in opening up your airway.

Many patients prefer oral appliances due to their size, comfort, and convenience. Some of the most common reasons why CPAP poses difficulty for patients are:

- bulkiness
- unattractive appearance
- bed partner does not like it
- patients are unable to sleep on their back
- cuddling becomes difficult
- maintenance is required
- extension cords are needed when staying in hotels
- patients are embarrassed to use when sharing a room
- a claustrophobic feeling

Patients can't believe that such a small device can give them a quality of life that they gave up on through the use of the CPAP therapy. The task of finding a dentist that performs oral appliance therapy is often difficult for many patients and typically requires them to have to depend on the Internet to find someone. But how reliable is this?

Physicians suggest that patients speak to their dentist, but many dentists are not well trained in providing services related to dental sleep medicine. Many times, your dentist may attempt to talk you out of wearing such devices because of the side effects. However, the only problem is that they are misinformed or do not understand what it takes to limit such negative effects.

By finding a dentist that specializes in dental sleep medicine, you can gain further insight into sleep apnea and proper treatments. The availability of oral appliances for the treatment of snoring and sleep apnea not only improves your overall health but also improves your personal life. So when it comes to oral appliance therapy, what appliance works best? Let's take a closer look at oral appliances and what we have to offer.

You've decided to undergo oral appliance therapy, but which one is the best option?

It is my goal to help each of my patients get a better night's sleep this year, and every year. For patients suffering from sleep apnea, an individualized treatment plan is available to meet their needs.

The American Academy of Sleep Medicine (AASM) has endorsed oral appliance therapy as an acceptable treatment modality for snoring and sleep apnea. The AASM also published a practice parameter paper in 2006 discussing the utilization of oral appliances in mild to moderate cases as a first line choice and a second line choice in severe sleep apnea cases for those that are noncompliant with CPAP therapy.

Currently, the largest number of patients suffering from obstructive sleep apnea are in the mild to moderate categories. Patients in these categories can be treated with oral appliances as the first line of treatment.

Despite the fact that treatment with the CPAP unit is extremely successful, some patients cannot, or choose not to, wear the facemask with the attached air compressor. For these patients, oral appliances provided by a dentist trained in sleep disorders are proving to be an excellent treatment option.

Once you have made the decision to utilize oral appliance therapy in the treatment of sleep apnea, life only gets easier. In general, all oral appliances perform similar functions. Typically, variations in the design and mechanism of holding the jaw forward are what differentiate each oral appliance. The idea of treatment is to have an oral appliance that is comfortable and effective in reducing or improving your sleep apnea.

Presently, there are well over forty different oral appliances that one could have custom-fabricated for the treatment of snoring or sleep apnea. Typically, the dentist selects the device based on his or her experience with such devices. This is no different from prescription medications where physicians get comfortable writing specific drugs, yet there are many that they can choose from for a particular condition.

With several oral appliances available, it is important to understand which devices will best meet your sleep apnea needs. To better understand the treatment options available for sleep apnea, let's take a closer look at a few of the oral appliances offered so that you can get a better night's sleep every day of the year, not just one. For continued understanding of the oral appliance therapy options available to you, please refer to the illustrations and brief descriptions of each oral appliance therapy option available below.

The Narval Appliance

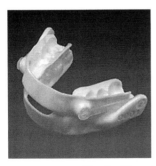

We will first discuss the Narval CC oral appliance—a new player in the field of dental sleep medicine. What is unique and interesting about this device is that it is printed on a 3D printer. Through this process your dentist will make a mold of your teeth and bite in order to allow your jaw to come forward by a certain percentage. Typically, about 50 to 70 percent of your ability to bring your jaw forward will be captured using dental putty. The molds are scanned into a computer and a technician designs the device. For this device to work, you will need to have teeth that are not short, and the side walls toward the cheek and tongue need to not be flat. This allows the appliance to grab onto your teeth so that it does not slip out while you are asleep. The Narval CC oral appliance will be printed out in resin, which allows the device to be thin, light, and extremely flexible for extreme comfort.

The Narval oral appliance is discreet and comfortable, utilizing technology that has been developed and refined for over a decade. Instead of pushing the lower jaw forward, the Narval appliance holds it in the forward position, minimizing the stress on the joint and limiting the risk of dental side effects. It provides greater comfort and flexibility with a slim design that is adjustable, has less stress on the front teeth, and is not a one-size-fits-all appliance.

What Is the Narval CC?

As we stated previously, this device is a custom-made mandibular repositioning device (MRD). Most MRDs hold the lower jaw in a forward position, but the Narval CC works along the occlusal plane to retain the mandible in a protruded position rather than pushing it. By doing this, it relieves the stress on the temporomandibular joint (TMJ). Through this elevated articulation point, the connectors are able to be parallel with the jaw line, which helps to compliment the physiological articulation, making the device more comfortable for your patients.

A Look at the Benefits

ResMed's Narval CC is known for innovation and excellence when it comes to treating sleep apnea. It is also the first and only CAD/CAM mandibular repositioning device. To help you better understand the benefits of this innovative device, here are three reasons why this MRD is the best choice in the treatment of sleep apnea and snoring:

1. Innovation—By being able to utilize CAD/CAM for this device, it helps to ensure a strong and precise fit that is custom-made to each patient's specific needs.
2. Size—With this device, you don't have to worry about bulk and discomfort. Instead, the Narval CC is thin and flexible, making it one of the lightest devices on the market today.
3. Comfort—Your patients need comfort when it comes to treatment, which is why the Narval CC is the best choice. With Narval CC, we optimize articulation along the occlusal plane to provide superior patient comfort, so it won't interrupt their sleep.

The Thornton Adjustable Positioner (TAP)

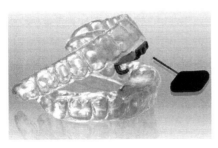

The Thornton Adjustable Positioner (TAP) is another series of appliances made out of triple laminate or ThermAcryl plastic. Besides the Narval, all other appliances currently available are made out of acrylic. This device has mechanics in front of the appliance that allow the lower jaw to be held forward. The TAP appliances are one of the first devices of its kind to enter the market. It is a versatile sleep apnea appliance that is connected by a hook located at the front of the upper appliance.

The TAP was created with the same principle as that of cardiopulmonary resuscitation (CPR). Your airway must be opened to allow for air to properly pass through the throat. If you have a constricted or collapsed airway, it causes snoring or sleep apnea. With the TAP device, it holds the lower jaw in a forward position so that it does not fall open during the night and cause the airway to collapse, maintaining a clear airway to reduce snoring and improve breathing.

With this device, you don't have to worry about discomfort. As it is custom-made, the TAP device is comfortable and helps to prevent any change in teeth position or mouth structure with the use of the AM Aligner. It is also patient adjustable, which further allows for maximum comfort and effectiveness in treatment.

The Dorsal Appliance

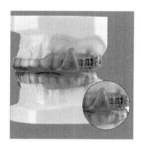

Several different laboratories make the next oral appliance: the Dorsal-type appliance. The difference with this device is the advancement mechanism and placement of it. The basic design and idea of the Dorsal appliance is a finlike concept, which resembles the dorsal fin of a shark. The fin engages in the form of a block that prevents the lower jaw from falling back during sleep. The common ones are Somnomed, Respire blue series, Dorsal appliance from Dynaflex, and the MAP-D from MAP Laboratories.

The original design of the Dorsal appliance dates back to the mid-1980s and has been one of the most popular sleep apnea appliances. The Dorsal appliance is a two-piece construction that allows for comfort and lateral jaw movement. The Dorsal fins on the appliance create a specific mandibular position. This oral appliance can be created in a variety of materials, including acrylic, dual laminate, or thermal splint material.

The Klearway™

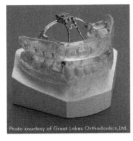

Photo courtesy of Great Lakes Orthodontics, Ltd.

Another appliance is the Klearway™, which was invented in 1995. This is an appliance fabricated from thermo-active acrylic (when warmed, the device softens and once cooled it hardens), which helps in easily fitting the appliance over your teeth. This device is one piece consisting of an upper and lower aspect. The Klearway is today's most thoroughly researched oral appliance for the treatment of snoring and mild obstructive sleep apnea (OSA).

It is clinically proven and provides the following special features for improved treatment and care:

- extensively researched and tested
- multiple positions available
- patients can easily adjust the device
- comfortable and well tolerated by patients

Extensive research for the Klearway oral appliance was funded by the Canadian government and contained clinical trials and crossover studies that compared effectiveness with other oral appliances and CPAP. Through this research, Klearway has been proven superior to other oral appliances for the treatment of snoring and sleep apnea due to its thermal sensitive material for a better fit, comfort and a better retention.

Herbst Appliance

If your face type is characterized by a recessed lower jaw, it can be assumed that you would have an overbite. Interestingly, this face type may also increase your risk of having sleep apnea. One orthodontic approach would be to have you wear full braces while having a Herbst device attached to your braces. What this does is it brings your lower jaw forward and helps you grow into your new bite.

The Herbst-type mechanics have been utilized in various oral appliances to help bring your lower jaw forward, leading to an ideal bite. It allows us to bring your lower jaw forward while you sleep in an effort to keep your airway open and prevent it from collapsing. We know that with sleep apnea, your jaw position and bite may play an important role, which is why the Herbst device is so beneficial in treatment. When your jaw is not properly aligned, your tongue can fall to the back of your mouth, which results in symptoms of sleep apnea.

There are many different Herbst-type appliances currently available on the market. Through a proper consultation with your dentist, the best type of oral appliance will be chosen for your treatment plan. Examples of various Herbst type devices are the Respire Pink series, SUAD (strong upper airway dilator), UCLA Herbst, MAP-H, and what some just refer to as a Herbst appliance.

The SUAD and TRD Oral Appliances at a Glance

 Strong upper airway dilator (SUAD) appliance is a standard Herbst-design appliance as described above. What makes this appliance different is that it has a metal framework that is cast to fit your arch shape. This helps to make the device stronger from the standpoint of breakage. It is a premium dental sleep appliance that is developed for the treatment of snoring and obstructive sleep apnea, and it is an effective, comfortable, and durable alternative to CPAP therapy or surgery.

By wearing the SUAD device while sleeping, your lower jaw will be moved forward into a comfortable position, which allows relaxation of the tissues at the back of your throat. This also ensures that the base of your tongue does not collapse and block your airway, giving you a safe and soundless sleep.

Tongue Retaining Devices (TRD)

One of the first devices utilized in management of snoring and sleep apnea, a tongue retaining device works on the simple principle of holding your tongue forward, and had no significant component of moving the lower jaw forward. Picture yourself sticking your tongue out at someone and your tongue being held in that position when you sleep. Presently, its use is limited, but at times it is utilized in cases where one has no teeth and wears dentures, or someone experiencing significant jaw pain is unable to move the tongue forward. In such instances, a TRD would make the best option for treatment of sleep apnea.

The Elastomeric Mandibular Advancement (EMA)

Elastomeric mandibular advancement (EMA) is yet another simple device that has an upper and lower component similar to the thickness of bleaching trays but a bit firmer. On the side, there are hooks where various sizes and thicknesses of elastics are placed to assist in advancing the lower jaw.

The Elastic mandibular advancement (EMA) oral appliance is customizable and removable. This appliance is created for the noninvasive treatment of snoring and sleep apnea. It is designed to advance the mandible while also opening your bite to allow for less restricted airflow during sleep. The use of EMA appliances helps to promote a deeper, more restful sleep by preventing snoring and relieving symptoms of sleep apnea. With a proper consultation, we can determine if an EMA appliance is appropriate for your individual case.

My Airway Positioner (MAP-T)

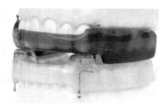

My Airway Positioner (MAP-T) is also a new addition to the choice of oral appliances and is quite unique in its own way. The MAP-T is a simple and comfortable device similar to the other appliances, with the exception that the mechanism that advances the jaw is placed on the biting side of the appliance. This allows almost no irritation to the cheeks or tongue since the mechanism to move the jaw forward is set away from such structures. The device is fully adjustable so that the lower jaw can be positioned forward or backward based on your individual needs.

I have had the privilege of working with many custom sleep apnea appliances throughout my career. Through the use of oral appliances, I have also learned directly from my patients' experiences, some of whom have expressed the various difficulties experienced during treatment. As in any industry or situation, you have individuals that want to improve a product to address many of the common issues that people express. Well, I was one of them. With that being said, I enlisted the help of a very good friend of mine that I highly respect because of his generosity but also his creativity.

Hector Rico is a certified dental technician that I have had the honor of working with for many years in appliance fabrication for my TMJ patients. Hector and I spoke at great lengths on current appliances on the market and what could be done to make them better, as well as address patient concerns. With many prototypes and me being the test subject, or what you may call a very courageous guinea pig, we started to develop what is known today as the MAP-T appliance. After many variations, materials, placement of mechanics, and months of sleepless nights for both him and me, we finally got to the bottom of many of the issues that patients would express.

As a result, the MAP-T, as described earlier in the chapter, was born. We started to introduce this appliance to our patients and started to note any issues that still existed. We were happy to share that the majority of concerns our patients had with previous devices were addressed. While needs and technology advancements continue to expand each year, we understand the need for updating our appliances. With the help and creativity of Mr. Rico, we will continue to improve the MAP-T appliance so that it continues to provide the best results for patients suffering from sleep apnea.

As someone who suffers from sleep apnea myself, I have worn pretty much all of the devices that are presently on the market. With that being said, don't let my good looks fool you, but this is a typical reason why people don't think they have such a life-threatening condition—you do not have to be overweight to fit the stereotype of having sleep apnea. I have worn all of the devices that I presently use on my patients at my practice, and I am proud and impressed at how simple, comfortable, and effective the MAP-T device really is.

Over-the-Counter (OTC) Boil and Bites

I always get asked about the over-the-counter (OTC) boil and bite sleep appliances that are advertised to help with snoring. As you may realize, the majority of advertisements will never say that it is used for sleep apnea since now you are engaging in the treatment of a medical condition. The marketing efforts for the OTC are to target the population that snores and not those who suffer from sleep apnea.

By reading this book you will have a better understanding that while snoring is a potential sign of sleep apnea, it does not always mean you have sleep apnea, and vice versa. The irony in this is that most patients with sleep apnea will snore. Because of this, a consumer may get an OTC device to help with snoring, but this may not address their sleep apnea (which many patients may not even realize they have). In my practice, testing patients with an OTC device many times worsens their undiagnosed sleep apnea.

Moving back to my patients asking about boil and bite over the counter appliances, we can understand how these devices do not offer any solution to their symptoms. My stand on these devices is to try it with caution. If you can tolerate it and see a reduction in your snoring, then consider a custom oral appliance as a treatment option once you have been diagnosed. However, let's first make sure what we are dealing with—simply snoring or some level of sleep apnea. Based on your diagnosis, we can properly formulate a treatment path. The end result will be a better night's sleep and a happier sleeping partner.

Dual Therapy

There are times when the standard therapy of utilizing CPAP is difficult due to the pressure setting that is required to help with your sleep apnea. In these cases, using an oral appliance with your CPAP will help. What you actually are facilitating is moving some of the obstruction forward with the oral appliance while also readjusting your CPAP pressure (in a lab setting) to determine the adequate setting.

This allows you to get the benefit of CPAP therapy with a much more tolerable pressure setting helping with your compliance. We have many patients that may use both their oral appliance and CPAP at home, but when traveling, they may only use their oral appliance. As described earlier in the book, there are many reasons why patients do not like to travel with their CPAP machine, and the use of an oral appliance provides them with an opportunity to be therapeutic, even for a short period of time over their travel experience. You also have to understand that for the patients currently not in this situation, something is better than not doing anything at all.

There are also ways to attach a nasal pillow mask to an oral appliance. An example is a product such as a CPAP PRO. By using a CPAP PRO, which is basically nasal pillows, attached to a fork that is attached to an oral appliance, you eliminate the need for any straps that are required to keep a traditional mask on your face. The oral appliance not only will bring the lower jaw forward but also will support the CPAP nasal pillows, allowing a much lower pressure setting. Through this, it helps to make it more comfortable and tolerable.

The role of a dental practitioner in the treatment of sleep apnea is important. While sleep specialists are routinely in charge of diagnosis and treatment selection, dental practitioners are best equipped to treat sleep apnea with oral appliances also known as mandibular repositioning devices (MRDs). We as dentists are available as your partners to help treat sleep apnea and snoring symptoms so that you can experience a better night's sleep.

Temporary Oral Appliances

I understand that sometimes you just don't want to wait for the creation of your appliance. Instead, you want to start treatment immediately while waiting for your permanent device. Well, you are in luck. There are a couple of temporary oral appliance options that will provide you with immediate treatment of your sleep apnea. The first one we will look at is myTap.

In a category of its own, myTap is the only precision-fit oral appliance that is both patient friendly and clinician friendly. It is also one of the quickest and most effective snoring and sleep apnea treatments available, based on the TAP technology. Currently, the myTap has been clinically approved in thirty-two independent peer-reviewed studies.

Combined with the patented TAP technology, this allows the patient to control their own therapy with sleek remoldable trays. These trays allow the appliance to fit effortlessly on the teeth while instantly creating a comfortable, low-profile fit. Since the myTap device can be reheated and refitted, it eliminates the need for follow-up appointments because it gives the patient the ability to achieve an ideal treatment position every time. While the myTap is not a custom-made appliance, it is customized at the time of fitting.

Another option is the Silent Sleep™ Oral Appliance. This oral appliance is available to provide relief for an array of conditions and is currently used to treat snoring, obstructive sleep apnea, and bruxism. It is also used:

- to test a patient's response to oral appliance therapy
- temporarily while a custom oral appliance is created as a spare appliance while traveling
- as an alternative to CPAP while traveling
- as a long-term appliance when custom appliances are not an option

- on patients with orthodontics
- as a sports mouth guard

The Silent Sleep™ oral appliance utilizes a one-piece tray system that is customized to the patient's teeth using GC Reline™, which is an in-office material (vinyl polysiloxane). It is fit to the posterior teeth while the patient protrudes their lower jaw slightly, which helps to maintain a patent oropharyngeal airway to reduce snoring and/or sleep apnea. With the Silent Sleep™ oral appliance, patients don't have to worry about treatment being too expensive, so they can receive the treatment they need right away.

To determine the best oral appliance, or temporary oral appliance, for sleep apnea treatment, it is important to schedule an appointment with a dental sleep medicine expert. With a proper consultation, you can find relief from your symptoms so that you can get a better night's sleep—every night.

Case Studies That Showcase Successful Oral Appliance Therapy

To help you better understand how important sleep apnea treatment is, and just how beneficial oral appliance therapy is, let's take a look at three different cases. I have had the opportunity to work with so many different patients in the treatment of their sleep apnea, and because of that, I want to share with you three cases that further showcase just how beneficial oral appliance therapy can be. Each case experienced different symptoms and utilized different treatment plans but experienced the same result: a better night's sleep and improved overall health and happiness.

(For the purpose of including the following case studies, patients' names have been changed for confidentiality reasons.)

Case #1

Meet Helen. Helen visited my office because she lives in a neighboring state and found us on the Internet. She came to my office experiencing the following symptoms:

- morning head pain
- headaches
- migraines
- jaw pain
- facial pain
- neck pain
- ringing in the ears
- dizziness
- fatigue
- pain when chewing
- nocturnal teeth grinding (bruxism)
- jaw clicking and locking

Each of these symptoms led to Helen feeling exhausted and unhappy. From the symptoms she experienced, you can see why she might be so exhausted and had grown irritable as the days went on. By reaching out to my office, Helen was hoping that we would have a solution that would help her get a better night's sleep, while also living pain free. And she was right.

Before reaching out to us, Helen was taking Cambia, Cataflam, Cymbalta, Flexeril, Levothyroxine, and Neurontin. She has significant medical history that, as you learned in previous chapters, could be the reason why she developed sleep apnea and other symptoms. To name a few, her past medical conditions included acid reflux, prior orthodontic therapy that involved removal of all her first premolars, and wisdom teeth removal. Currently, however, Helen was experiencing anxiety, chronic pain, difficulty speaking, dizziness, nasal allergies, psychiatric care, sinus problems, sleep apnea, and thyroid disorder.

She also previously had various surgeries, including jaw surgery in which her jaw was moved forward and titanium rods were placed. Her family history also included diabetes and high blood pressure, while her mother currently has chronic headaches, poor sleep, snores, and high blood pressure. Through these indications, it should have been clear that sleep apnea was a possibility.

In addition to her list of symptoms and medical history, Helen went through a long list of unsuccessful treatments. She received treatment from a neurologist for migraine pain with Botox. She also visited a sleep medicine specialist for treatment of sleep apnea with CPAP and also received a dental device. Helen even visited a chiropractor for treatment of her TMJ pain for routine neck and back adjustments. As you can see, Helen experienced an array of symptoms, medical history, and treatments from various specialists but still suffered from her array of symptoms.

Upon visiting my office, Helen expressed that her pain had progressively gotten worse, especially while she had braces. The pain was so severe that she was unable to identify the origin. When asked if anything made her pain or discomfort worse, she stated, "I wake up and it never resolves. Migraines make the facial and TMJ pain worse, and it is constant." Since she was diagnosed with sleep apnea in 2012, Helen has worn a Herbst-type device that aggravated her TMJ symptoms. Presently, it is not treating her sleep apnea, and she cannot tolerate CPAP therapy.

Helen was experiencing tenderness in her neck on the right side, as well as her face muscles responsible for closing the mouth and chewing. She was able to open her mouth to about twenty-nine millimeters without pain, but when she opened her mouth at a maximum of thirty-eight millimeters, she experienced significant pain.

We decided to fit her with a MAP-T appliance. (We described this oral appliance earlier in the chapter.) At her first follow-up visit, Helen stated she was sleeping better but snored a little. Because of her facial pain, we decided to advance the lower jaw slowly. With this slow, advanced approach, Helen visited us at her next follow-up after six weeks and did not express any jaw pain, face pain, pain when chewing, or any fatigue. In the past, she had braces and also had four teeth extracted (first premolars) in addition to her wisdom teeth to make room to straighten her teeth; there was not much room for her tongue, which may be why she had sleep apnea in the first place.

We then sent her back to her sleep doctor in her hometown and they repeated a sleep study with the appliance in her mouth. Through this, we found that her AHI was down to 3 and her minimum oxygen level was at 94 percent. Before oral appliance therapy with the MAP-T, Helen's baseline study showed an AHI level of 19.3 and a minimum oxygen level of 79 percent.

Case #2

Tanya is a thirty-seven-year-old female with a BMI of 27 (not overweight by any means). She would complain to her primary care doctor that she was tired and moody all the time. Her husband suggested that she see her medical doctor to get medications because she would always have a short fuse with him and the kids. Her doctor had her on antidepressants with some success, but she was still always tired. After complaining to him over multiple appointments, he referred her to get a sleep study.

With a history of sleep apnea, she was prescribed CPAP treatment, but this offered no success. While wearing the CPAP machine, she felt that someone was choking her and had a difficult time breathing with it on. With this treatment, her baseline sleep study showed her to have an AHI of 31 and a minimum blood oxygen level of 85 percent. Other than depression, Tanya had no health issues.

She was then referred to us from her primary care physician. Through a detailed examination of her dental structures, we created her a custom Dorsal-type appliance. After this, she would return almost weekly complaining of irritation to her cheeks and difficulty falling asleep because of the amount of material in her mouth. In the past, she had braces and also had four extra teeth extracted (first premolars) in addition to her wisdom teeth to make room to straighten her teeth up; there was not much room for her tongue, which may be why she had sleep apnea in the first place.

After her complaints, we decided to switch her to a MAP-T appliance, and wow!, was she happy! She came into our office for her follow-up visit with a cake and a smile on her face—we are always thrilled when our patients come in so happy. Her husband was even grateful because her mood was different around him and their children. With the use of the MAP-T appliance, Tanya was finally sleeping and resting. Can you imagine if someone kept waking you up because you were not breathing (arousals described earlier)? In Tanya's case, she was waking up at least thirty-one times a night, even though she was unaware of it. If this was the case for you, you would wake up "cranky," as well.

The utilization of the MAP-T appliance allowed Tanya to properly sleep, breathe, and wake up refreshed. Her family was even able to witness this change in her mood and behavior, which improved their overall home life. Her follow-up study with the oral appliance that was performed in the sleep lab showed her AHI drop from a 31 down to a 4.6 and her minimum blood oxygen level went up to a 95 percent.

Case #3

A forty-four-year-old man, Rick had a BMI of 29 and traveled for work. He came into our office because he was tired of taking his CPAP machine with him. Many times he would purposely leave his CPAP machine at home when he would travel. By doing this, he would be tired at the clients' sites and unable to concentrate and function as he needed to. This started to affect his job performance and he needed to change that. His medical history included borderline diabetes and he was asked to control it by exercise and diet.

He was diagnosed with having sleep apnea and had been faithful in wearing the CPAP machine when at home, but not while traveling. In addition to his sleep apnea, Rick was a teeth grinder, which meant he would wear a night guard for it. He destroys his night guards every couple of months due to the grinding he does and has to get them fixed or replaced frequently.

Rick has a base line dx of AHI 10 with a minimum oxygen level of 85 percent. We were able to fit him with a Narval CC appliance made by ResMed. The advantage of using this device was that it is the most durable and strongest device on the market with a three-year warranty so that his aggressive teeth grinding would not ruin it. And if his grinding was to, the warranty covered its replacement up to three years.

At his first follow-up visit to our office, he was impressed on how comfortable the Narval sleep apnea device was in comparison to his night guard. Through this, he was also not snoring at home with his oral appliance and could tell a difference in how he felt. We had him advance the device a few more times and even placed him back in the lab for further testing. At this point, his AHI was at a 3.2 and his minimum blood oxygen was at 92 percent.

Rick continues to wear his oral appliance all the time. Since he needed to wear a night guard with the CPAP machine, he decided to only wear the Narval oral appliance because it serves both of his needs. With the Narval device, his teeth are protected and his airway was opened up, which meant he got a better night's sleep.

A Rare Occurrence in Successful Treatment

We have previously touched base on three cases of patients suffering from sleep apnea. These cases showcased how our patients went through numerous treatments and studies without success. But by visiting our office for oral appliance therapy, they were able to find the results they needed to sleep better at night, while living a healthy, happy life. To expand on these cases, I was introduced to a patient recently that suffered from sleep apnea, but the end result was different than most.

I had a patient that had been sleeping in a different room from his wife for over six years because of his snoring. He was diagnosed with having moderate sleep apnea, but the snoring was way too much for his wife to handle. Through research, his wife found our office and sent her husband to us for diagnosis and treatment options.

He came into our office to see what we could offer him for treatment of his sleep apnea. After discussing his sleep apnea, we created an oral appliance for him. When he came back in for his follow-up appointment, he was not happy. When we asked him why he was not happy, he explained that he was not snoring, he feels energetic, he was able to concentrate, and he no longer had any urge to take naps in the late afternoon like before. So why was he unhappy?

This patient was unhappy because he has to sleep with his wife, who hogs and pulls the bed covers, which wakes him up every night. While we successfully treated his snoring and sleep apnea through oral appliance therapy, we are not able to fix bed cover thieves.

Chapter 10

Other Procedures

Procedures for mild to moderate sleep apnea include uvulopalatopharyngoplasty (nicknamed UPPP or U triple P), genioplasty, hyoid suspension, tongue suspension, and different forms of midline tongue resection. For moderate to severe sleep apnea, the list includes forms of jaw repositioning, hypoglossal nerve stimulators, and tracheostomy. Weight-loss surgery falls into a different category entirely and, for those who are morbidly obese, this may be a very viable option.

Uvulopalatopharyngoplasty

By far the most commonly performed of these procedures is the uvulopalatopharyngoplasty. The U triple P has been around for more than fifty years. The basic idea behind it is that the palate rattles when you snore and contributes to sleep apnea. It's typically performed alone or in conjunction with a tonsillectomy. The very back part of the roof of the mouth is soft, and the little dangling thing that hangs down in the middle is called the uvula. The very back part of this soft tissue is taken off so that it doesn't collapse into the back wall of the throat as much. This is about as painful as a tonsillectomy on steroids during an ice-cream shortage. Aside from being sadistically painful, our biggest beef with the procedure is that it simply does not work all that well. In fairness, it does help some people, but only if they have mild to moderate sleep apnea. Mild to moderate sleep apnea, as you will recall, is fewer than thirty episodes per hour of airway collapse. The typical success rate for U triple P is about a 50 percent reduction in sleep apnea. Because it is such a painful procedure, we typically reserve it as one of the last resorts for our patients.

One of the most difficult challenges for a surgeon is to determine, or rather anticipate, the degree to which scarring will impact the final configuration of any tissues being worked on. One of the worst problems with the uvulopalatopharyngoplasty procedure is that a surgeon can occasionally—not usually but occasionally—cause nasopharyngeal stenosis. Nasopharyngeal stenosis is a condition wherein the back part of the nose is tightened from the contracture of the scar and breathing is compromised. To make matters worse, the unfortunate part is that this problem is very, very difficult to correct.

The most common long-term side effect of the uvulopalatopharyngoplasty procedure is difficulty swallowing. Normally, difficulty with swallowing after this procedure, due to the pain, lasts for three weeks. But one of the possible outcomes of having the scar contract is that the palate, or roof of the mouth, can shift forward so much that you can no longer close off the nose when you swallow. When this happens, food or liquid tends to go in the nose and may be difficult to remove.

Remember what we are trying to correct here is to keep the airway from collapsing and keep our patients from having oxygen deprivation at night. Remember also that oxygen deprivation raises the risk of stroke and heart attack and triggers a host of other problems, including the exacerbation of diabetes, etc. So it's not that we feel that the uvulopalatopharyngoplasty has no benefit but that it must be prescribed judiciously, only after less invasive options have been explored.

Genioplasty

Another surgical option, the genioplasty, has become fairly uncommon in recent years. The idea for this procedure is to take one of the anterior attachments of the tongue to the jaw and move it forward. The genial tubercle is a small point in the front of the jaw just below the lower two incisors. Small squares are cut into the lower segment of the jaw and it is shifted forward. That pulls one of the muscles of the tongue and the hyoid forward together. The problem is it doesn't shift the tongue forward that much, and over time, the muscles of the tongue relax backward again. Our further assessment concludes that the procedure requires a fair amount of experience and has a relatively high initial failure rate. However, the impact of that failure rate is buffered by a fairly small complication rate as well as the inconspicuous location of the surgical scar beneath the chin.

Hyoid Suspension

The hyoid suspension is a procedure with which we have similar misgivings. What we've seen of this procedure is that it tends to fold the tongue backward onto the top of the voice box. Moreover, in terms of problems with swallowing and pain, it tends to result in a fair amount of *both* postoperatively. It has been our experience that both problems tend to last for an extended period of time. Due to what we have concluded to be the limited success of this procedure, we don't use it very often. From our perspective, the procedure is classified as "possibly helpful." Ironically, it remains one of the standard procedures that is paid for by insurance companies. Go figure.

Tongue Suspensions

Tongue suspensions use a suture material to anchor the tongue to the front of the jaw. This is typically from behind the front lower edge of the jaw. An anchor (typically some form of screw or other metallic device through the lower portion of the jaw in the midline) is attached to a suture which lassos the tongue. This is then tightened to keep the tongue from falling all the way to the back of the throat while the patient is lying flat. Postoperative swelling is manageable but a bit of a problem and the patient needs to be watched overnight. The recovery is typically about two weeks.

Again, there are some problems swallowing postoperatively. The swallowing problems can last a fairly long time but generally settle down several weeks after the procedure. Bleeding is occasionally a problem because there are two fairly large arteries just off the midline on either side close to where the sutures are placed. The biggest problem with the procedure is that over time the suture tends to cut its way through the tissues, much the same way as a cheese cutter works through a block of cheddar. As a result, the tongue eventually relaxes backward during sleep yet again. This can take anywhere from a year to five years. In terms of where this procedure is placed in the decision tree, we typically use it as a procedure of second or third line resort after the nose has been addressed and the patient has failed treatment with the VOAT procedure.

Midline Tongue Resections

Midline tongue resections were originally described decades ago. They have come back into vogue with some of the new technology that allows very precise visualization of the tongue with imaging and excellent bleeding control. Like the VOAT, this procedure is performed in the midline to avoid nerves and blood vessels. Recently, we have seen some pretty good successes for this particular procedure. Ultrasonic Doppler devices and preoperative imaging with CT's and endoscopy have allowed for very precise procedures in the last several years.

In particular, there is a new device out which causes a plasma at the tip of the instrument wand which basically evaporates a part of the tongue. This allows for very controlled removal of tissue. As you can imagine, this procedure is very painful and requires postoperative nutritional support in the form of IV fluids and/or tube feedings. Some potential long-term swallowing problems may also arise. For these reasons, this procedure is also viewed by our practice as a second or third line procedure.

Jaw Repositioning

For more severe sleep apnea, there are jaw-positioning procedures. Jaw repositioning is typically performed by oral maxillofacial surgeons. It is very effective in treating sleep apnea. However, the surgery requires cutting at least the lower jaw, and sometimes the upper and lower jaws and repositioning portions of the face. As you can imagine, with the larger procedures come greater risks of complications. Getting the dental bite just right is a challenge. Therefore, it is best to choose an experienced surgeon. It is possible to create unintended changes in the shape of the nose, changes in the shape of the face, and damage to the nerves that move the face and jaw. Nasal obstruction may result, as well as chronic sinusitis. The hardware that is used to reposition the face can become painful or infected. When repositioning the upper jaw, there's even the possibility of causing eye damage, although this is rare. Some studies have suggested that the results may not be permanent and the sleep apnea tends to come back after five or more years.

Because of these considerations and the prolonged recovery, this procedure is viewed by our practice as a second or more likely third line treatment option. There is one exception to this rule in that for patients who have retrognathia, a condition characterized by extremely small jaws, this lower jaw repositioning can in fact be very helpful.

Hypoglossal Nerve Stimulation

A relatively new procedure for sleep apnea is hypoglossal nerve stimulation. The stimulator is wrapped around one of the hyperglossal nerves nearest the tongue base with the intention of stimulating the tongue in order to keep it out of the airway at night. We have seen some patients who experienced initial success, but over time, they no longer seemed to benefit from the procedure. Our conclusion is that the procedure is so new that adequate data on success rate is not yet sufficient to make a reasonable judgment on where this procedure fits in the treatment scheme.

Tracheostomy

The final procedures done specifically to address sleep apnea is the tracheostomy. The tracheostomy is a hole in the neck that simply bypasses the tongue, nose, and all the structures above the windpipe. It definitively fixes the sleep apnea, but there are a lot of problems with tracheostomies. To begin with, the tracheostomies for sleep apnea patients are frequently done only for morbidly obese patients. This means that there are a lot more tissues to go through to get from the surface of the skin to the inside of the windpipe, and bleeding can become more of a problem. It also means that the tube has to be longer, has to have more room to move around, and thus might more likely become displaced. The longer tubes also tend to become clogged with mucus more easily. Mucus is the slime that's produced inside your nose and inside of your windpipe. When it clogs up the windpipe, it can be deadly. In fact, for long-term tracheostomy patients, there is a much increased mortality rate. This means that patients with tracheostomies tend to plug their way or bleed into the airway much more frequently, and often die from this complication alone. A tracheostomy tube can erode into some major vessels within the surrounding area and cause catastrophic bleeding. That is not to say that tracheostomies are not a better option for these patients. Without it, their apnea-induced low oxygen would aggravate heart problems, etc. The procedure just needs to be used with extreme caution. Tracheotomy is definitely a third line option.

Chapter 11

VOAT and OAT: Combination Therapy

Great things are done by a series of
small things brought together.

—Vincent Van Gogh

To obtain better results, we offer a combination therapy of ventral-only ablation of the tongue base (VOAT) and oral appliance therapy (OAT). As combination therapy, these two treatments offer the best of both worlds. Both VOAT and oral appliance therapy by themselves help to reduce AHI. However, in the end, the combination of the two therapies helps patients receive superior results comparable to outcomes to wearing a CPAP (without the CPAP).

The VOAT approach is to use radio frequency ablation (RFA) to shrink the tongue, which has greatly improved since past treatments of such procedure. Previous RFA procedures made small incisions on the top surface of the tongue while the VOAT procedure makes incisions on the bottom surface, where scarring is less visible, less painful, and patients have an 80 percent chance of improving. In fact, the VOAT procedure is quick, requires minimal sedation, and can sometimes last only thirty seconds. The end result is a procedure that has become one of the best options for potentially curing sleep apnea.

Oral appliance therapy (OAT) is available as an alternative to CPAP in the treatment of sleep apnea. Currently, many people who are noncompliant to CPAP visit their ENT for treatment options, but it has also been shown that your dentist can be a part of the first line of defense against sleep apnea through the availability of OAT. Today, the largest number of patients who are suffering from obstructive sleep apnea are those in the mild and moderate categories. Your dentist will work with you to determine which oral appliance would work best for your individual needs and symptoms.

While VOAT and OAT work on their own to help in the treatment of sleep apnea, the two treatments together provide even better results. The combination of the two treatments gives patients the better of both worlds and has been proven to be better than wearing a CPAP, a therapy to which many patients are noncompliant. The combination of VOAT and OAT affords patients the same therapeutic results as CPAP, while avoiding the compliance issues and side effects inherent with CPAP therapy entirely.

To better showcase the benefit of dual treatment with VOAT and OAT, let's take a look at Steven. Steven had a pre-AHI of 34 and was fitted with an oral appliance. After a few weeks of advancing the device, he felt better and was not snoring. However, he was still not well rested in the morning. A home sleep test was performed to determine why Steve felt he was not feeling rested. His study showed that his AHI was reduced to 11 with the oral appliance. Steven was then referred to Dr. Dillard because he was not able to advance the sleep appliance devise anymore. Dr. Dillard performed two VOAT procedures a month apart. A month after Steven's last procedure, his AHI was down to a 3. His remaining complaint of not feeling rested was resolved. Steven continues to wear his oral appliance and feels great.

With the combination of the VOAT procedures and OAT, Steven was able to find relief from his sleep apnea symptoms so that he could get a better night's sleep. VOAT and OAT allow patients to find relief from their symptoms for a better night's sleep that produces better results than with CPAP alone.

Ideally, it makes sense that both VOAT and OAT have a level of success rate when performed independently. In a large number of cases to get the most out of non-CPAP related treatment, a combination of both VOAT and OAT would allow success rates comparable to one wearing CPAP alone, again with fewer side effects. So if you are looking for freedom from CPAP, consider combination of VOAT and OAT to obtain the results you need.

Chapter 12

Self Help

If you don't take care of yourself, the undertaker
will overtake that responsibility for you.

—Carrie Latet

One of the least complicated and most beneficial things you can do
to help with your snoring or to improve your sleep apnea is lifestyle
modification. This common-sense approach can be as simple as losing
weight or avoiding the use of tranquilizers as sleep aids. Additional
healthy choices that can greatly impact your snoring and/ or sleep
apnea include avoiding alcohol for at least four hours and heavy meals
or snacks for three hours before you go to bed. Establish regular
sleeping patterns and sleep on your side rather than your back. If
you have allergies, treat them. Please take these steps to improve the
quality of you sleep, as well as your overall health.

Body Length Pillow

Sleep apnea, and for that matter the occurrence of snoring, is more
prevalent in patients who sleep on their back. While we realize

that orthopedists, chiropractors, and even yoga instructors might champion the benefits of decompressing the spine by sleeping on your back, our primary focus is on making sure that your brain and body receive a steady flow of oxygen. We feel strongly that this basic need trumps any other concern and has a greater impact on your overall health.

With that said, we sometimes suggest that patients with low to moderate levels of sleep apnea who habitually sleep on their back try a full-length body pillow. These pillows relieve pressure points that can make adjusting to sleeping on your side more tolerable for some patients. Since it has been demonstrated that gravity can effectively cause the tongue to close the narrow but critical path of oxygen in patients who sleep on their back, trying to sleep on your side is a common-sense approach and definitely fits our criteria of being less invasive and having few—pardon the pun—side effects. Body pillows are available from retail stores and online sleep supply vendors. We recommend shopping for the pillow that feels right for you. Like a mattress, or even a standard pillow, relative softness, thickness, and shape retention all affect comfort in ways unique to the individual. We suggest that you try out the various types of body pillows to give yourself the best possible opportunity to improve your sleep and reduce the occurrence of snoring and quite possibly the frequency of any sleep apnea episodes which you might be experiencing. We remind you that if you are experiencing symptoms of sleep apnea, it is advised that you be tested to determine the severity of your condition. Once diagnosed, you are encouraged to look at all treatment options available to manage your condition.

Sleep: The Link between Longevity and Health

A good night's rest is the key to longevity, health, and beauty. We might find ourselves skipping an hour or two of sleep just so we can finish work or get that extra time to spend with friends or family, but in doing so, we are really only hurting ourselves. Research consistently shows that people who get both quality and quantity shuteye are less susceptible to serious illness like heart disease, high blood pressure, diabetes, and obesity. Let's again take a closer look at sleep in relation to longevity and health.

After just one day of poor sleep, many of us feel the effects the next day. We may feel drowsy, our memory doesn't work as well as we would like, and we seem to struggle to perform even routine mental and physical tasks. However, we rarely take into account the long-term effects and the impact on our longevity because these are far more subtle and we have become accustomed to sleeping poorly. The more sleep we lose, the more pounds we pack on and the more we suffer from common colds and the flu without connecting our sleep habits to the increased frequency of these ailments.

Sleep quality has been linked to weight and our immune system for some time now. Research continues to show that inadequate sleep significantly alters the metabolic pathways that regulate appetite, which ends up making us feel hungrier than we really are. A lack of sleep also causes us to get sick easily. While many of us may not think that missing an hour or two of sleep is detrimental to our health, that misconception can prove dangerous. Sleep directly relates to both longevity and health, making it imperative that we get the sleep that we need. If we struggle with a sleep disorder, we are not letting our bodies do their best work, which puts us at risk for more serious illness that impacts our longevity.

There is a common saying that no one will take up for you as much as you will. While the inherent truth of this statement might be brought into question, it is important that each of us is willing to advocate for his or her own heath care. With that said, it is important to understand that do-it-yourself health care is generally not the best policy. There are a lot of things you can do for yourself, and it is important to always remember that among them is your decision to seek treatment and subsequently your level of advocacy for the quality of care your receive from your doctor.

Self-help is not limited to the untreated. Sometimes, patients can benefit by making small adjustment to a therapy put in place by their doctor. Such common-sense adjustment can go a long way toward ensuring therapeutic success. Let's begin with the nose. One of things that happens when people try using CPAP is that problems with the nose can develop. The first thing you can do is look in the mirror and try to breathe in and decide whether not your nose is collapsing at the opening. If this is happening, you may want to consider using Breathe Right strips or a similar product to help you decide whether not this simple solution can improve your sleep. If you find relief, the next step would be to look online, and search for alternatives to breathe right strips or nasal valve collapse devices for a long-term, more cost-effective solution that would cause less potential damage to your skin over time.

For best outcomes to be obtained with the Breath Right strips, it is essential to find the point at which the nose is collapsing the most. By marking the spot with a small dot of ink, you can make certain that the strip is not positioned too far out onto the face to be effective. If the properly applied strips succeed in keeping the nasal airway clear, Dr. Dillard suggest trying a fairly innovative device which is simply an elastic plastic cone that goes inside the nose. His wife calls this the "cone of silence," which boomers reading this book might recognize as a reference to *Get Smart*—the Don Adams original. (Matthew Broderick may have also used the device; somehow we missed that movie.)

It's also always helpful to have someone around to tell you whether or not this morning was better than yesterday morning or whether or not you appear to be breathing better when you are trying out these different things. In the event that you sleep alone, it can be useful to try to use some sort of a recording device. Be sure you keep the recording device at the same distance from the bed before and after you try any new treatment.

A measure of effectiveness that you can record without the use of technology is whether or not you have increased or decreased sleepiness or fatigue. A useful tool for this is the Epworth sleepiness scale. We've included in chapter 2 for your convenience, and it's useful to keep a tracking sheet to check your progress. A diary or journal can also be utilized effectively for this scenario. These self-analyses are somewhat inherently subjective but fairly reproducible. It is not a substitute for a sleep study or a doctor.

Another thing to track your progress is to check and see whether or not your blood pressure or blood sugar is elevated more or less after your new adaptations. Of course, this presumes that you have either diabetes or high blood pressure and are monitoring your own levels.

Back to our list, the second thing for the nose is to try and make sure that the nose is well humidified. You can do this with a variety of over-the-counter medications. The key elements of this is to remember that the nose itself has to make mucus, and sometimes it can fall behind. Especially in an era when people are very accustomed to treating their noses with Neti pot, nasal irrigations, or nasal saline, it is very frequent to see people who have excessively irrigated their nose and damaged the lining by using these techniques.

One piece of evidence that you have overdone it with the irrigation is to see small amounts of blood with the irrigation or to have nosebleeds. In these cases, it's very useful to use a nasal emollient. However, if you are not improving with a nasal emollient after one or two days, you should seek medical attention.

You should also see an ENT if you're having more than three or four sinus infections lasting more than a week over the course of one year, if you have lost your sense of smell, or if you are persistently experiencing discolored drainage from the nose.

Ponaris, NeilMed nasal spray gel, Ayr gel, any nasal spray with aloe vera gel, and Yerba Santa are preferred emollients. Alternately, you might choose to make an effective and low-cost emollient at home. My "solution" combines saline and baking soda with one tablespoon of corn syrup per liter. I've included a detailed recipe below.

- 8 ounces of boiled water, allowed to cool
- One half teaspoon of baking soda (not baking powder)
- One teaspoon of non-iodized salt (iodine irritates the nose)
- One tablespoon of white corn syrup

Boil the squirt bottle in water in order to clean it. The solution can be kept in the refrigerator for two days. At night, just before bed, you can place a clean cotton tip applicator coated with saline in each nostril. Coat the inside of each nostril liberally.

Another thing you can do to get a handle on your sinus situation is the Afrin challenge test. Again, take note of what your Epworth sleepiness score is and what your snoring is like prior to doing this. If you have significant health problems, such as heart arrhythmias, heart conditions of any sort, thyroid problems, or severe blood pressure problems, you should check with your doctor prior to beginning this test. Also, please note that you should not continue to use this drug more than two to three days without first checking with your physician. To execute the test is pretty simple: two squirts in each nostril about an hour and a half to two hours before bedtime, and see if you sleep better.

Once you've completed the Afrin challenge test and determined whether or not you have seen improvement, how do you interpret the results? Usually, if you have experienced noticeable improvement, a procedure inside the nose will probably give you an improvement in your breathing.

Again, one of the simplest things you can do is to stop eating late. The old saying that extols the virtue of eating breakfast like a king, lunch like a prince, and dinner like a pauper applies. One of the most clinically effective simple truths of the saying is the encouragement to eat very little at night. One of our observations is that people with sleep apnea almost always have a lot of acid reflux. In our office, we test for reflux on a regular basis. We have multiple patients who have had the reflux fixed and find that their sleep apnea improved after the reflux procedure. Common sense tells us that cutting down on the amount of food that you eat before you go to bed tends to reduce the amount of reflux that you experience at night. There's a pretty good bet that it will help your sleep as well. How long you should not eat before you go to bed is a question that no one can really answer with any degree of accuracy. Our best-guess recommendation is that you probably should not eat for four or five hours before you go to bed.

Along the same lines, you probably should not eat very spicy foods or drink alcohol, carbonated drinks, or juices before you go to bed if you have sleep apnea. It's possible that taking over-the-counter acid-reducing medications can help the situation as well. Although prolonged use of these medications should be cleared with a physician, a short trial may help a number of different things, including sleep apnea.

It's really helpful to elevate the head of your bed as well. The ideal angle is probably the same angle you get out of a recliner. Keeping your head elevated seems to help a number of things, including improved positioning of the tongue. While not immediately obvious, the tongue tends to hang back into the throat and occlude it more when you are lying flat. In addition, lying back makes it easier for acid to run up from the stomach into the airway.

Another element of positioning is that the side position seems to be a lot more advantageous for mediating sleep apnea. A number of garments have been sold in the past which had balls sewn into the back of the shirt discouraging lying flat on your back. Honestly, this is not such a bad idea. As any husband can attest, you get fewer complaints and a lot fewer elbows to the ribs when you lie on your side.

Exercise seems to have a beneficial effect, even without the benefit of weight loss. Of course, lowering your caloric intake and trying to exercise at the same time have multiple benefits far beyond just helping to deal with sleep apnea.

If you already have a CPAP machine, then you should continue to try to use the machine until you are officially cured of sleep apnea. One of the things that can help you tolerate CPAP much better, for those of you like to sleep on the side, is a CPAP pillow. CPAP pillows are specially constructed with a small hole in the pillow, which the CPAP fits neatly into. Another nice accessory device is a foam wedge which will help elevate the head, and it can be used to help keep you on your side.

Conclusion

Getting a good night's sleep is not only measured by how much you sleep but also by how well you sleep. Most readers are likely already aware that the deepest phase of sleep is referred to as the REM phase. What you may not realize is the extent to which REM is your brain's unsung hero. It is during REM sleep that the most intense neural activity occurs, signaling increases in blood circulation and oxygen levels. During this phase, our brain tissue also absorbs more amino acids. REM is the reason why good sleepers are not only sharper thinkers but are also less at risk for neurological diseases later on in life, such as Alzheimer's disease.

This book has presented information about sleep disorders, specifically obstructional sleep apnea, and the serious health risks associated with disordered sleep. Our goal has been to inform patients about the necessity of quality restorative sleep and to provide hope and encouragement through our discussion of treatment options, strategies for adapting to therapies, and common-sense practices designed to help you sleep better.

By arming yourself with information, you will discover that there really is hope. We now end this book with a reminder that the danger of inaction when symptoms of sleep apnea are present is too great to take a wait-and-see approach. Your sleep apnea will not improve without treatment.

We have presented both surgical and nonsurgical techniques for the treatment of sleep apnea. It is important for both you and your doctor to realize that neither sleep apnea nor its treatment can be addressed with a "one-size-fits-all" label or solution. Whatever your current condition, treatment history, past failures, frustration, or apathy toward your sleep disorder might be, we again urge you to seek treatment or to advocate for better care. There's a solution out there that can provide the lasting relief and life-changing therapy that you need in order to start living with renewed zeal and vigor. The solution is more likely than not a painless and noninvasive therapy that will provide you with lasting relief. So set aside your fears. Take charge of your health. The next move is yours. Contact a sleep specialist in your area as soon as possible. The decision to take action—to address and treat your symptoms of sleep apnea—will indeed dramatically improve, and quite possibly save, your life.

Remember sleep apnea *will hurt you*. Thanks to advances in modern medicine, the cure no longer has to.

References

1. Woodson BT, Steward DL, Weaver EM, Javaheri S. "A Randomized Trial of Temperature Controlled Radiofrequency, Continuous Positive Airway Pressure, and Placebo for Obstructive Sleep Apnea Syndrome." *Otol He3N Surg* 128(6): 848–61, 2003.
2. Steward DL, Weaver EM, Woodson BT. "Multilevel Temperature-Controlled Radiofrequency for Obstructive Sleep Apnea: Extended Follow-Up." *Otol He3N Surg* 132(4): 630–5, 2005.
3. Woodson BT, Nelson L, Mickelson SA, Huntley T, Sher A. "A Multi-Institutional Study of Radiofrequency Volumetric Tissue Reduction for OSAS." *Otol He3N Surg.* 125(4): 303–311, 2001.
4. Li KK, Powell NB, Riley RW, Guilleminault C. "Temperature-Controlled Radiofrequency Tongue Base Reduction for Sleep-Disordered Breathing: Long-Term Outcomes." *Otol He3N Surg* 127(3): 230–4, 2002.
5. Friedman M, Ibrahim H, Lee g, Joseph NJ. "Combined Uvulopalatopharyngoplasty and Radiofrequency Tongue Base Reduction for Treatment of Obstructive Sleep Apnea/Hypopnea Syndrome." *Otol He3N Surg* 129(6): 611–21, 2003.
6. Kezirian EJ, Powell NB, Riley RW, Hester JE. "Incidence of Complications in Radiofrequency Treatment of the Upper Airway." *Laryngoscope* 115(7): 1,298–304, 2005.

7. Steward DL. "Effectiveness of Multilevel (Tongue and Palate) Radiofrequency Tissue Ablation for Patients with Obstructive Sleep Apnea Syndrome." *Laryngoscope* 114(12): 2,073–84, 2004.

8. Steward DL, Weaver EM, Woodson BT. A Comparison of Radiofrequency Treatment Schemes for Obstructive Sleep Apnea Syndrome." *Oto He&N Surg* 130(5): 579–85, 2004.

9. Stuck BA, Kopke J, Hormann K, Verse T, Eckert A, Bran G, Duber C, Maurer JT. "Volumetric Tissue Reduction in Radiofrequency Surgery of the Tongue Base." *Oto He&N Surg* 132(1): 132–5, 2005.

10. Fischer Y, Khan M, Mann WJ. "Multilevel Temperature-Controlled Radiofrequency Therapy of Soft Palate, Base of Tongue, and Tonsils in Adults with Obstructive Sleep Apnea." 113(10): 1,786–91, 2003.

11. Pazos G, Mair EA. "Complications of Radiofrequency Ablation in the Treatment of Sleep-Disordered Breathing." *Otol He&N Surg* 125(5): 462–6, 2001.

12. Robinson S, Lewis R, Norton A, McPeake S. "Ultrasound-Guided Radiofrequency Submucosal Tongue-Base Excision for Sleep Apnoea: A Preliminary Report." *Clin Otolaryngol* 28(4): 341–5, 2003.

13. Stuck BA, Kopke J, Maurer JT, Verse T, Eckert A, Bran G, Duber C, Hormann K. Lesion Formation in Radiofrequency Surgery of the Tongue Base." *Laryngoscope* 113(9): 1,572–6, 2003.

14. Troell RJ. "Radiofrequency Techniques in the Treatment of Sleep Disordered Breathing," *Otol Clin N Am* 36: 473–493, 2003.

15. Powell NB, Riley RW, Guilleminault C. "Radiofrequency Tongue Base Reduction in Sleep-Disordered Breathing: A Pilot Study." *Otol He&N Surg* 120(5): 656–64, 1999.

16. Riley RW, Powell NB, Li KK, Weaver EM, Guilleminault C. "An Adjunctive Method of Radiofrequency Volumetric Tissue Reduction of the Tongue for OSAS." *Otol He&N Surg* 129(1): 37–42, 2003.

17. Ceylan K, Emir H, Kizilkaya Z, Tastan E, Yavanoglu A, Uzunkulaoglu H, Samim E, Felek SA. "First-Choice Treatment in Mild to Moderate Obstructive Sleep Apnea: Single-Stage, Multilevel, Temperature-Controlled Radiofrequency Tissue Volume Reduction or Nasal Continuous Positive Airway Pressure." *Arch Otolaryngol Head Neck Surg.* 2009 Sep; 135(9): 915–9.

18. Fernández-Julián E, Muñoz N, Achiques MT, García-Pérez MA, Orts M, Marco J. "Randomized Study Comparing Two Tongue Base Surgeries for Moderate to Severe Obstructive Sleep Apnea Syndrome." *Otolaryngol Head Neck Surg.* 2009 Jun; 140(6): 917–23.

19. Eun YG, Kwon KH, Shin SY, Lee KH, Byun JY, Kim SW. "Multilevel Surgery in Patients with Rapid Eye Movement-Related Obstructive Sleep Apnea." *Otolaryngol Head Neck Surg.* 2009 Apr; 140(4): 536–41.

20. Farrar J, Ryan J, Oliver E, Gillespie MB. "Radiofrequency Ablation for the Treatment of Obstructive Sleep Apnea: A Meta-Analysis." *Laryngoscope.* 2008 Oct; 118(10): 1,878–83.

21. Eun YG, Kim SW, Kwon KH, Byun JY, Lee KH. "Single-Session Radiofrequency Tongue Base Reduction Combined with Uvulopalatopharyngoplasty for Obstructive Sleep Apnea Syndrome." *Eur Arch Otorhinolaryngol.* 2008 Dec; 265(12): 1,495–500.

22. van den Broek E, Richard W, van Tinteren H, de Vries N. "UPPP Combined with Radiofrequency Thermotherapy of the Tongue Base for the Treatment of Obstructive Sleep Apnea Syndrome." *Eur Arch Otorhinolaryngol.* 2008 Nov; 265(11): 1,361–5.

23. Nelson LM, Barrera JE. "High Energy Single Session Radiofrequency Tongue Treatment in Obstructive Sleep Apnea Surgery." *Otolaryngol Head Neck Surg.* 2007 Dec; 137(6): 883–8.

24. Friedman M, Lin HC, Gurpinar B, Joseph NJ. "Minimally Invasive Single-Stage Multilevel Treatment for Obstructive Sleep Apnea/Hypopnea Syndrome." *Laryngoscope.* 2007 Oct; 117(10): 1,859–63.

About the Authors

Dr. David G. Dillard

Dr. David G. Dillard, a Navy veteran, is board certified by the American Board of Otolaryngology, Head and Neck Surgery in Otology (ear diseases and surgery), Rhinology (nose and sinus diseases and surgery), Laryngology (voice box disorders and surgery), Head and Neck Surgery (primarily cancer surgery of the throat, nose, larynx and neck) and Facial Plastic Surgery. He has practices in both Athens and Lawrenceville, Georgia.

Dr. Dillard has over a decade of experience in the treatment of ENT problems with special interests in sinusitis, ear problems, vertigo, and obstructive sleep apnea. In fact, Dr. Dillard is the patent-pending innovator of multiple devices and treatments that are the result of his many years experience in his field. In regards to sleep apnea, he has performed over 500 of his patent-pending V.O.A.T. surgeries. Board Certified, Dr. Dillard held an academic appointment at Emory University School of Medicine, currently holds an academic appointment at the University of Georgia School of Medicine, and has received a national award from the American Academy of Otolaryngology, Head and Neck Surgery.

Dr. Dillard has published a number of research articles in major Otolaryngology journals. In addition, Dr. Dillard has also been awarded research grants and has given presentations at a number of national academic meetings. Dr. Dillard is a graduate of the Emory University School of Medicine, Department of Otolaryngology residency. In addition to providing treatment that can potentially cure your sleep apnea, Dr. Dillard also treats sinusitis and other conditions of the nose and sinuses. In fact, Dr. David G. Dillard is one of the most experienced Sinuplasty surgeons in the country. He was the first surgeon certified in Georgia for in-office Sinuplasty.

His expertise on Balloon Sinuplasty has been featured in Sky Magazine, Atlanta Medicine, The Journal of the Medical Association of Atlanta, as well as on radio stations WGKA and WAFS.

Dr. Mayoor Patel

Dr. Mayoor Patel is a dentist with advanced knowledge and experience in dental sleep medicine and orofacial pain. He received his dental degree from the University of Tennessee in 1994, and went on to complete a one-year residency in Advanced Education in General Dentistry (AEGD). In 2011, Dr. Patel earned a Masters in Science from Tufts University in the area of Craniofacial Pain and Dental Sleep Medicine – an area of dentistry that continues to grow each year.

Currently, Dr. Patel serves as a board member with the Georgia Association of Sleep Professionals, the American Board of Craniofacial Dental Sleep Medicine, American Board of Craniofacial Pain and American Academy of Craniofacial Pain. He has also taken on the role as examination chair for the American Board of Craniofacial Pain and American Board of Craniofacial Dental Sleep Medicine.

Through his commitment and dedication to a variety of associations, Dr. Patel strives to help educate other dentists in the area of Craniofacial Pain and Dental Sleep Medicine. He understands that not many dentists are aware of the need for experts in this extensive area of dentistry, which is why he not only continues to educate himself, but other dentists, as well. Dr. Patel's expertise and knowledge in Craniofacial Pain and Dental Sleep Medicine places him in a category of itself, which puts him at a significant advantage.

With extensive knowledge and expertise, Dr. Mayoor Patel has also served as Director of Dental Sleep Medicine for FusionSleep from 2008-2014 and as Adjunct Faculty Member at Tufts University from 2011-2014. Currently, Dr. Patel is an Adjunct Faculty member with Georgia Regents University, The Atlanta School of Sleep Medicine and clinical director of education for Nierman Practice Management.

Since 2003, Dr. Patel has limited his dental practice to the treatment of TMJ Disorders, Headaches, Facial Pain and Sleep Apnea – areas he constantly lectures on and consults with other dental practices to help implement. Expanding on his expertise and commitment to education, Dr. Patel has also made additional contributions that have been published in textbook chapters, a consumer book on treatment options for sleep apnea and several journal articles. He even holds patents on an oral appliance to help successfully manage obstructive sleep apnea so his patients can get a better night's sleep.

Dr. Mayoor Patel continues to lecture both locally and nationally to the dental and medical communities in the area of Craniofacial Pain and Dental Sleep Medicine. His expertise in these areas makes him a valuable asset to the Dental Sleep Medicine community to help dentists better understand the treatment of these conditions within their own dental practice.